BRITAI
DRAGON

A field guide to the damselflies and dragonflies of Britain and Ireland

Dave Smallshire & Andy Swash

WILD *Guides*

PRINCETON
press.princeton.edu

Published by Princeton University Press,
41 William Street, Princeton, New Jersey 08540
In the United Kingdom: Princeton University Press, 6 Oxford Street,
Woodstock, Oxfordshire OX20 1TW
nathist.press.princeton.edu

Requests for permission to reproduce material from this work should be sent to
Permissions, Princeton University Press

First published 2004 by **WILD**_Guides_ Ltd.
Second Edition 2010 by **WILD**_Guides_ Ltd.
Third Edition 2014

British Library Cataloging-in-Publication Data is available

Library of Congress Control Number 2013952020
ISBN 978-0-691-16123-5

Production and design by **WILD**_Guides_ Ltd., Old Basing, Hampshire UK.
Printed in China

10 9 8 7 6 5 4 3 2 1

Contents

THE SPECIES ACCOUNTS

Foreword

Dragonflies are Rock-and-Roll insects. They are high-energy aerial predators that have all the raptorial appeal of birds of prey, the grace, agility and vibrancy of butterflies plus, for those with patience enough to sit and watch the water on a still and sunny summer's day (and, let's face it, what better place to be and what better occupation?), they have the 'watchability' that is second to none in the insect world. In some cases, the dramas and excitement that unfold are so special that, I can honestly say, in some instances even compete with some of my best birdwatching moments!

As insects, the Dragonflies and Damsels have it all. They get pretty much everywhere: every freshwater body, whether a stagnant moorland bog pool, a 'lily-padded' lake, duck pond, canal, river, or even a garden pond made from an old bath-tub, can support these gauzy-winged lovelies.

Their accessibility goes even further than this. Since I was a kid, I have reared countless 'pond dragons' in aquaria on my kitchen table, and watched in awe as the stealthy nymphs murdered and mashed everything from water-fleas to sticklebacks with their formidable face-gear. Then on 'D' day (that's 'D' for Dragonfly you understand), been spell-bound as these cryptic uglies have crawled from their aquatic nursery and burst there and then in front of my very eyes to become the perfect vision of winged beauty.

Identification is the first step in understanding any animal and it adds a value to moments in the field if you know what you are looking at. Within the dragonflies and damselflies there are enough distinct species not to deter the amateur, but there are also enough surprises, some tricky species and a handful of migrants and variations to keep the 'experts' on their toes and retain that wonderful sense of mystery that hooks us all.

It is with these points that this book comes into its own. It will become my very own Dragon-hunting manual and will hopefully shed as much light on these insects in the field for you as it certainly will for me.

Introduction

Dragonflies are stunning and amazing insects! They are often very brightly coloured, kill for a living and have phenomenal powers of sight, flight and manoeuvrability. Some are large, though none as big as the one-metre wingspan giant Protodonata that flew some 325 million years ago. These are considered to be the ancestors of Dragonflies, which were well in evidence during the heyday of the dinosaurs. Almost 6,000 species are recognised today, but only a fraction of these have ever been seen in Britain or Ireland, and only 40 or so currently have breeding populations.

Dragonflies are characterised by having an aquatic larval stage, incomplete metamorphosis, two pairs of wings and large, compound eyes. The wings that have proved so important over the aeons are incredibly light and yet very strong. Powered by large muscles in the thorax, they enable larger species to travel at up to 36 km/h. They also allow all species to hover if they so wish. Their multi-faceted eyes provide excellent colour vision and acuity. Add to this a death-trap of a food-collecting 'basket' of legs and razor-sharp mandibles, and you have quite a fearsome beast! Fortunately, despite appearances, Dragonflies are harmless to humans.

Dragonfly populations throughout the world are highly dynamic and these are exciting times for the Dragonfly-watcher! Whether it is because of their brilliant colours, predatory habits, association with water, or the challenge of identification, dragonflies have grabbed the attention of an increasing army of fans.

The compound eyes of a Brilliant Emerald, which make for a formidable hunter.

The wings of this newly-emerged Four-spotted Chaser show the complex structure that gives them great strength – a feature that has served Dragonflies and their predecessors well for aeons.

About this book

This book aims to provide the tools for anyone interested in damselflies and dragonflies to improve their knowledge and enjoyment of these incredible insects, and to contribute towards their conservation. The focus is on the identification of both adult forms and larvae. It is illustrated throughout with carefully selected, high-quality photographs and detailed and accurate illustrations of key features. Introductory sections provide a summary of Dragonfly biology and ecology to help you better understand and interpret what you see in the field, and tips on how, when and where to look. The book covers all 56 damselfly and dragonfly species that have ever been recorded in Britain or Ireland, as well as a few that might possibly turn up as vagrants. Individual species accounts make up the bulk of the book, each highlighting the key identification features of adults. These accounts also provide up-to-date information on the species' status,

The term Dragonfly (with a capital 'D') is used in this book for the Order Odonata, which includes both dragonflies (or Anisoptera, meaning unequal wings) and damselflies (Zygoptera, meaning equal wings). Throughout this book blue text is used to denote damselflies (Zygoptera) and green text for dragonflies (Anisoptera). The following photographs depict a typical damselfly and dragonfly, and highlight their distinguishing features.

Key features of an adult damselfly	Key features of an adult dragonfly

Blue-tailed Damselfly	*Golden-ringed Dragonfly*
Damselflies are distinguished by:	**Dragonflies are distinguished by:**
Typically small size and dainty proportions	Robust form
Flight usually weak and brief	Flight powerful, often persistent and hovering
Wings usually held together over back	Wings held at right angles to the body at rest
Front and hind wings identical in shape	Front and hind wings have different shapes
Eyes separated	Eyes touch in most species

behaviour, habitat preferences, distribution and flight periods. Towards the back of the book is a separate, detailed section on the identification of larvae. This has been carefully structured and illustrated with field identification in mind. Technical terms have been avoided as far as possible; those used are explained in a Glossary (*page 58*).

Many Dragonfly enthusiasts and most newcomers to the subject use English names in preference to scientific names. For this reason, English names are used throughout the text. When referring to a particular species these are shown with initial capitals (*e.g.* Southern Hawker); references to groups of species are shown in lower case text (*e.g.* hawkers). For those who prefer to use scientific names, these are included in the main species accounts. The English names of British species follow those recommended by the British Dragonfly Society (BDS), while the names of vagrants and species not yet recorded in Britain follow those in Dijkstra and Lewington (2006); other commonly used names and, where they differ, those used by Nelson and Thompson (2004) for Ireland are provided in the main species accounts.

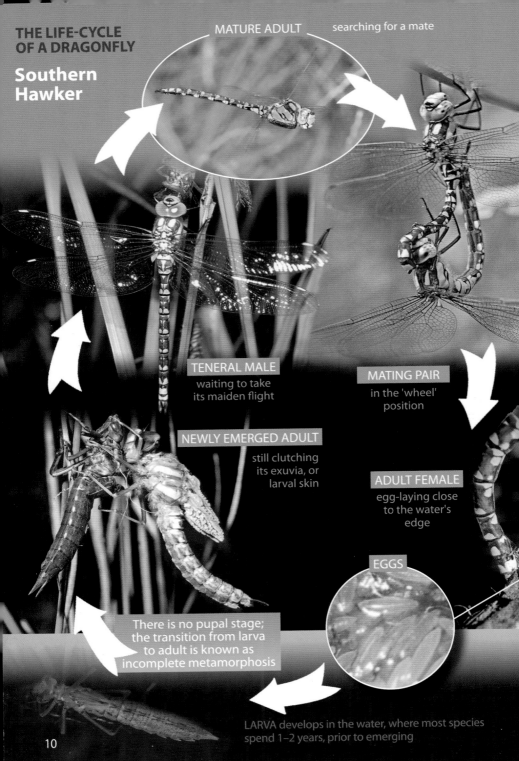

THE LIFE-CYCLE OF A DRAGONFLY

Southern Hawker

MATURE ADULT — searching for a mate

TENERAL MALE
waiting to take
its maiden flight

NEWLY EMERGED ADULT
still clutching
its exuvia, or
larval skin

MATING PAIR
in the 'wheel'
position

ADULT FEMALE
egg-laying close
to the water's
edge

EGGS

There is no pupal stage;
the transition from larva
to adult is known as
incomplete metamorphosis

LARVA develops in the water, where most species
spend 1–2 years, prior to emerging

Dragonfly biology and ecology

The illustration opposite summarises the life-cycle of a Dragonfly, and more detailed information on the egg, larval, emergence and adult stages is given in this section.

Egg

Female Dragonflies can lay hundreds of eggs during their adult lives, in batches over a few days or even weeks. Eggs are laid (oviposited) either into plant material (endophytic eggs) or deposited loosely into water (exophytic eggs). The former are elongated, but the latter are rounded and laid in a jelly-like substance which confers some protection.

All damselflies and the hawkers have scythe-like ovipositors and inject their eggs into plant stems or leaves, rotten wood or mud on or close to the surface of the water. Amazingly, some damselflies submerge completely to lay their eggs, often using their still-attached partner to pull them up again afterwards! Emerald damselflies and Migrant Hawkers inject their eggs into rush stems well above the water surface, while other hawkers lay into rotten wood or other debris just above the waterline. The spectacular Golden-ringed Dragonfly lays its eggs by hovering vertically and stabbing its abdomen into stream-beds. All other species, including the emerald dragonflies, chasers, skimmers and darters, repeatedly dip the tips of their abdomens into water, each time releasing one or more eggs that settle below the surface.

Endophytic eggs

Exophytic eggs

Southern Damselflies egg-laying 'in tandem', with the female completely submerged.

A pair of Large Red Damselflies egg-laying.
By remaining 'in tandem', the male prevents other males from mating with the female.

During egg-laying, male damselflies, chasers, skimmers and darters guard the females with which they have just mated, either by staying linked 'in tandem' or by flying close by. Female dragonflies often attract attention when they are egg-laying as their wings rustle against emergent plants.

Eggs hatch either within 2–5 weeks or, in the case of the emerald damselflies and some hawkers and darters, the following spring.

Larva (or nymph) (see *pp. 190–211*)

What emerges from the egg is tiny and tadpole-like, and designed to escape from the water. In cases where the egg is laid above the waterline, this 'prolarva' wriggles to safety as soon as it hatches, and moults within a few hours. Prolarvae emerging from eggs laid below the waterline moult almost immediately after hatching. During its time as a larva, the Dragonfly catches and eats live prey at every opportunity, moulting a further 5–14 times until it is fully grown.

A damselfly larva: Red-eyed Damselfly

Damselfly larvae are long and thin, with three leaf-like 'tails' (caudal lamellae) through which oxygen diffuses.

Larval development typically takes one or two years, but ranges from 2–3 months in the case of the emerald damselflies to more than five years in Golden-ringed Dragonfly. Development takes longer in cooler waters where food is scarcer, whilst in warm waters, such as those of southern Europe, there may be more than one generation per year. With global warming, two generations per year may become increasingly possible for some species in Britain and allow southern species, such as Red-veined Darter, to become firmly established. In any case, it is worth remembering that Dragonflies spend most of their lives underwater and only pass a few days or weeks in the adult form that interests most people!

All Dragonfly larvae have six legs (as in adults), wing-sheaths, a hinged jaw (labium) that can shoot out in an instant and catch prey, and the ability to breathe underwater. The labium is flattened in damselflies and hawkers, but spoon-shaped in other dragonflies. The appearance and habits of each species have evolved so that they can occupy different niches and avoid competition and predation.

The larvae of some species are covered in hairs that collect silt or organic debris from the soft sediment within which they live. Others are coloured green and/or brown, which helps to provide camouflage amongst the submerged plants and algae they inhabit. Species living in bottom sediments have relatively poor eyesight, but instead use their hairs, long legs and antennae to sense prey. In contrast, hawkers have large eyes and hunt by sight amongst plants nearer the surface.

A dragonfly larva: Southern Hawker

Dragonfly larvae take in water through their rectum into internal gills, and can be either torpedo-shaped or broader-bodied, flattened and spider-like. (They also have the ability to expel water forcibly when a burst of speed is needed!)

Prey includes insect larvae, crustaceans, worms, snails, leeches, tadpoles and small fish. Many species benefit from the annual abundance of frog, toad and newt tadpoles, which make relatively easy meals during late winter and spring. Water-fleas and midge larvae also provide abundant food sources, especially for damselflies and smaller dragonflies. Of course, Dragonfly larvae themselves can fall victim to predators, including other Dragonfly larvae, fish and waterfowl.

Permanent, unshaded and unpolluted waters with a rich variety of aquatic plants tend to support the greatest number and diversity of Dragonfly larvae. However, some species can tolerate brackish or polluted conditions and a few are adapted to living in new or temporary waters. Predation of larvae by fish can have a major impact on Dragonfly populations, perhaps explaining why acidic, fish-free waters are so good for Dragonflies.

Emergence

Emergence sequence of a dragonfly (Emperor Dragonfly)

1 The larva crawls out of the water and attaches itself to vegetation. **2** The adult then breaks slowly through its larval skin, the head and thorax exiting through a hole just in front of the wing sheaths. **3** There follows a resting period to allow the legs to harden, during which the adult hangs down (damselflies protrude forwards during this stage).

4 When the legs have hardened sufficiently, the adult reaches up, grabs its larval case (exuvia) and eases its abdomen free. **5** Body fluids are then redistributed and used to 'pump up' the wings and abdomen. **6** When fully expanded, the adult is about twice as long as its larval skin.

7

Shiny wings are characteristic of a newly emerged (teneral) adult.

7 After a further hour or so the wings, though still reflective, have hardened sufficiently for the teneral adult to take its maiden flight. This individual can be identified as a female Emperor Dragonfly from its abdominal patterning, but the identification of damselflies in particular is difficult until they start to colour up.

Unlike other winged insects, such as butterflies, Dragonflies do not have a pupal stage. Instead, the adults emerge directly from the larval skin during a final moult that takes place out of water. This is triggered by day length and temperature. In some species that emerge in spring, such as Emperor Dragonfly, emergence is synchronized with many individuals leaving the water on the same night. In others, the period of emergence is more extended.

Final-stage larvae sit in shallow water near the margins for several days, getting ready for their final moult and starting to breathe air. Most species leave the water during the morning, but hawkers often do so under cover of darkness. Larvae climb up robust emergent vegetation, a rock or sometimes an artifact. Some may walk several metres over dry land before finding a secure perch with enough space for the adult to emerge.

After finding a secure support, they redistribute their body fluids – first to push the thorax, head, legs and wings out of the larval skin. There is then a pause of about 30 minutes to allow their legs to harden enough for the next stage, when the abdomen is withdrawn. The wings and then the abdomen are expanded and start to harden. This process leaves behind the larval skin, called an exuvia, and lasts for between one hour (in the case of damselflies) and three hours (in the case of dragonflies).

Maiden flights are weak and typically cover only a few metres. However, in very warm conditions some species, such as Scarce Blue-tailed Damselfly, may rise into the air and drift for long distances at high altitude. During maiden flights, Dragonflies are especially vulnerable to predation by birds such as the Hobby. Smaller predators, like Blackbirds, spiders and ants, also take a heavy toll during emergence. Other hazards include physical obstructions or rainfall, which can result in many individuals suffering from imperfectly expanded wings or damage to the soft tissues.

Adult

Newly-emerged adult Dragonflies are known as tenerals. These are pale at first with only hints of the final adult patterning, have dull eyes and often whitish wing-spots (pterostigma – see *page 37*). At this stage, both sexes of many species appear very similar. Their wings remain reflective for a couple of days and the flight is weak and fluttery. As its body and wings harden, the adult begins to hunt for food. Tenerals spend about a week feeding away from water and gradually acquire adult colouration and sexual maturity. The maturation period depends on the species and the prevailing weather, lasting from a few days to a few weeks.

Tenerals, especially newly-emerged adults, are often very difficult to identify due to their lack of pigmentation. For this reason, the best features to look for are often size and structure, although the time of year may also help to eliminate some species.

Once mature, the adults move back to water to breed. The males of some species, notably dragonflies, are territorial and battle constantly to obtain and defend a suitable length of water's edge. They investigate any interloper and will try to seize any female in the hope of mating. The number of adults found at water is determined by the species' territorial behaviour. Territorial species are always found in smaller numbers than gregarious species, such as the blue damselflies, which may be present in hordes.

Ageing in Dragonflies (Scarce Chaser)

After hardening off, the teneral **t** has soft, reflective wings and reduced adult patterning. It attains female-like immature colouring **i** within a day or two.

As males reach sexual maturity, they go through a transitional phase **Mtr** during which the final mature colouration **Mm** is acquired.

In some species, such as Scarce Chaser, males develop 'pruinescence' – a waxy bloom that gives them a powder blue colour.

Colour changes are usually less marked in females **F**, though some damselflies go through distinct immature phases (*e.g.* the blue-tailed damselflies – see *pages 92–94*).

Older females **Fo** sometimes develop male-type colouration, though are never as bright as a mature male.

t

i

F

Mtr

Fo

Mm

Mating in Dragonflies is unique. The male grasps a female by the 'scruff of her neck' with the anal appendages (or 'claspers') at the tip of his abdomen: the pair is then said to be 'in tandem'. He may then transfer sperm from near the tip of his abdomen to accessory (or secondary) genitalia under the front (see *page 37*), although sometimes this happens before grasping a female. If the mood takes her, the female curls the tip of her abdomen to meet the male's accessory genitalia and sperm is transferred; this position is known as 'the wheel', or technically *in copula*. In some species, such as the chasers, the whole mating process takes only a few seconds; at the other extreme, Blue-tailed Damselflies take up to six hours. Lengthy periods of mating (more than 15 minutes) generally mean that the male has actively removed any sperm from any previous matings.

To some extent, pairing between different species is prevented by unique combinations of the shapes of the male's anal claspers and the rear of the female's head (in dragonflies) or pronotum (in damselflies). Although virtually impossible to see in the field, these can provide conclusive proof of identity in some difficult species (as can wing venation).

Egg-laying occurs soon after mating. Females that lay alone frequently have difficulty avoiding the attentions of males and therefore may choose to lay during poor weather, even in rain, or late in the day.

Adult Dragonflies are most active between mid-morning and mid-afternoon, when temperatures are highest. In Britain, flight is generally restricted to sunny weather, when the flight muscles in the thorax are sufficiently warm. This is especially the case for species that spend most of their time perched. Alternatively, a tactic used by hawkers, especially females (allowing them to lay eggs during cool weather), is to generate heat in the flight muscles by vibrating their wings. Generally, though, dragonflies seem to disappear in wet or cool cloudy conditions. In reality, they sit still and well hidden amongst vegetation, often hanging under leaves in trees.

The life-expectancy of adults is short, typically no more than a week or two, but exceptionally 6–8 weeks.

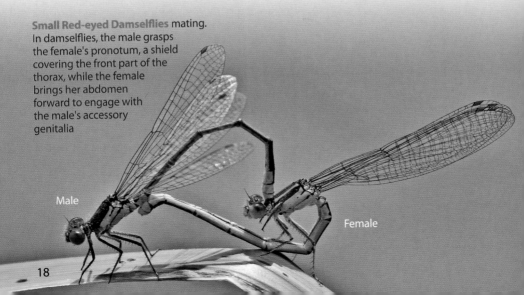

Small Red-eyed Damselflies mating. In damselflies, the male grasps the female's pronotum, a shield covering the front part of the thorax, while the female brings her abdomen forward to engage with the male's accessory genitalia

Male

Female

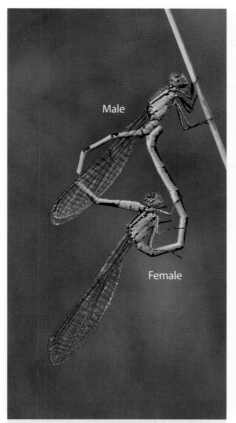

Male

Female

Common Blue Damselflies mating

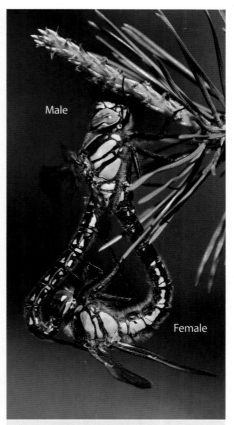

Male

Female

Hairy Dragonflies mating

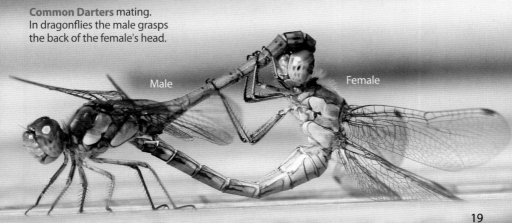

Common Darters mating.
In dragonflies the male grasps
the back of the female's head.

Male

Female

Dispersal and migration

Aquatic habitats may change in quality, sometimes quite rapidly. Dragonflies therefore need a mechanism for finding and colonising new breeding sites. Although their wings provide the means of achieving this, nearly all individuals and most species are quite sedentary. High population densities may drive dispersal, often during periods of hot weather and strong tail-winds.

Migrant Hawker: once regarded as mainly a regular migrant from mainland Europe, but now breeds widely across southern Britain.

Although most records of migrants have involved dragonflies, it is worth remembering that insects the size of a small damselfly flying higher than two metres are effectively in aerial circulation and at the mercy of prevailing winds. The extent to which dragonflies disperse in this way is uncertain, although it seems likely that most species fly (or are blown along) much closer to the ground. Studies of the highly localised Southern Damselfly have shown that most adults move less than 50 metres during their lifetime,

Yellow-winged Darter: a migrant that appeared in Britain in unprecedented numbers in August 1995.

although some travelled as far as two kilometres. However, it seems likely that on a hot day with a light wind, maiden flights of some species could take newly emerged tenerals high into the sky where they could potentially drift long distances. Indeed, at times of mass emergence, Hobbies may be seen catching tenerals over a wetland.

Some species, such as the demoiselles, White-legged Damselfly, Common Club-tail, Keeled Skimmer and Black Darter, regularly disperse a few kilometres. However, there are some well-established longer-distance migrants. Migrant Hawker is well named, being a regular immigrant along the south coast, although it also has a firmly established and expanding resident population. Other species whose numbers are regularly boosted by immigrants include Four-spotted Chaser and Common and Ruddy Darters.

Records of vagrants have increased in recent years, as interest and the skill levels of observers have grown. The species concerned include Common Green Darner (from North America), Vagrant Emperor (from north Africa) and Scarlet, Vagrant and Banded Darters (from mainland Europe). Other continental immigrants, such as Southern Emerald, Willow Emerald and Small Red-eyed Damselflies, Lesser Emperor and Yellow-winged and Red-veined Darters, have established breeding populations after arriving, although not all of these have persisted. It is clear that recent range expansions in Britain, the rest of Europe and elsewhere can be linked to climate change. On the other hand, the ranges of some northern species, such as White-faced Darter in England, have contracted further northwards.

Dragonfly habitats

The habitats favoured by breeding Dragonflies in Britain and Ireland can be broadly subdivided into eight categories: lake, pond, river, stream, canal, ditch, bog and flush. Each is illustrated in this section and the typical species that are likely to be found are listed. The tables on *pages 30–31* provide a summary of the favoured breeding habitats for all the breeding species.

Lake Typical species of both natural and artificial lakes include **Common Blue**, **Blue-tailed** and **Red-eyed Damselflies**, **hawkers**, **Black-tailed Skimmer** and **Common Darter**. Large, deep lakes and reservoirs generally support few species.

Pond Even small garden ponds can support **Large Red** and **Azure Damselflies**, **Southern Hawker** and **Common Darter**; in southern Britain they are often first colonised by **Broad-bodied Chaser**.

River **White-legged Damselfly**, **Common Club-tail** and **Scarce Chaser** are the specialities of some lowland rivers, occurring alongside more widespread species such as **Banded Demoiselle** and **Red-eyed Damselfly**, as well other common damselflies.

River Upland rivers hold a restricted range of Dragonflies but are favoured by **Beautiful Demoiselle** and **Golden-ringed Dragonfly**.

Stream Crockford Bridge, New Forest, Hampshire, is one of the best sites to see **Southern Damselfly**, with a typical wet heathland cast of **Beautiful Demoiselle**, **Scarce Blue-tailed Damselfly**, **Small Red Damselfly**, **Golden-ringed Dragonfly** and **Keeled Skimmer**.

Canal A wide range of Dragonflies can be found along well-vegetated canals, including localised species such as **Red-eyed Damselfly**, **Hairy Dragonfly** and **Downy** and **Brilliant Emeralds**. Infrequent boat traffic and good water quality provide the best conditions.

Ditch The best grazing marsh ditches hold **Variable Damselfly** and **Hairy Dragonfly**, with **Ruddy Darter** in those dominated by emergent vegetation. Some also support **Scarce Chaser** and, in East Anglia, they are the favoured habitat of **Scarce Emerald Damselfly** and **Norfolk Hawker.**

Flush Flushes often hold **Large Red Damselfly**, **Golden-ringed Dragonfly**, **Keeled Skimmer** and **Black Darter**, sometimes with **Small Red**, **Southern** and **Scarce Blue-tailed Damselflies** and, in Scotland, **Northern Emerald**.

Bog **Large Red**, **Small Red** and **Emerald Damselflies**, **Four-spotted Chaser** and **Black Darter** characterise bog pools in the lowlands. Upland bogs also support **Common Hawker**, and in Scotland **Azure Hawker**, **Northern Emerald** and **White-faced Darter** occur, sometimes at lower elevations.

Favoured breeding habitats

These tables summarise the main habitat preferences for all the Dragonfly species recorded breeding in Britain or Ireland. The key below explains the coding used to differentiate between common and widespread species, and those that are scarce.

SPECIES	HABITAT								Page no.
	Lake	Pond	River	Stream	Canal	Ditch	Bog	Flush	
Damselflies									
Beautiful Demoiselle			■						60
Banded Demoiselle			■						62
Emerald Damselfly		■							64
Scarce Emerald Damselfly	●	●				●			66
Southern Emerald Damselfly		●				●			68
Willow Emerald Damselfly	●	●	○	○		●			70
Large Red Damselfly		■					■		72
Small Red Damselfly		●		●		●	●	●	74
White-legged Damselfly	○	○	●	●	●				76
Southern Damselfly				●		●		●	78
Northern Damselfly	●	●					●		80
Irish Damselfly	●	●					○		82
Dainty Damselfly		●				●			84
Azure Damselfly	■	■			■	■			86
Variable Damselfly	●	●			●		●		88
Common Blue Damselfly	■	■	□	□	■	■	□		90
Blue-tailed Damselfly	■	■			■	■			92
Scarce Blue-tailed Damselfly		●		●				●	94
Red-eyed Damselfly	●	●	●			●			96
Small Red-eyed Damselfly	●						●		98

KEY: Common and widespread species: ■ Principal habitat □ Secondary habitat
Localised species: ● Principal habitat ○ Secondary habitat

Bog pools in lowland heathland Thursley Common, Surrey, is one of the best sites for dragonfly diversity in Britain, with 29 species recorded, including **Small Red Damselfly**, **Common Hawker**, **Downy** and **Brilliant Emerald**, **Keeled Skimmer** and **Black Darter** breeding in a range of acidic wetland habitats.

SPECIES	HABITAT								Page no.
	Lake	Pond	River	Stream	Canal	Ditch	Bog	Flush	
Dragonflies									
Hairy Dragonfly	●	●			●	●			100
Azure Hawker							●		102
Common Hawker		●							104
Migrant Hawker	●								106
Southern Migrant Hawker		●				●			108
Southern Hawker	●	●							110
Brown Hawker									112
Norfolk Hawker						●			114
Emperor Dragonfly	●								116
Lesser Emperor	●	●							118
Golden-ringed Dragonfly			●	●					120
Common Club-tail			●						122
Downy Emerald	●	●			●				124
Brilliant Emerald	●	●			●				126
Northern Emerald							●		128
Four-spotted Chaser		●					●		130
Broad-bodied Chaser		●							132
Scarce Chaser	●	●	●		●	●			134
Black-tailed Skimmer	●								136
Keeled Skimmer		●		●			●	●	138
White-faced Darter		●					●		140
Black Darter		●					●	●	142
Common Darter	●	●							144
Ruddy Darter						●			146
Red-veined Darter	●	●				●			148
Yellow-winged Darter	●	●					●		150

Where to look for rare, scarce and localised Dragonflies

The table opposite provides suggestions of sites to visit in order to see all the rare, scarce and localised Dragonfly species in Britain and Ireland. These sites are listed below, their approximate locations are shown on the map and their numbers are used in the table opposite. At least part of each of these sites can be accessed via public rights of way or are nature reserves where a permit may be obtained on site.

A note of caution
When searching for any of these species, be careful not to damage the habitat through trampling, or to cause undue disturbance to the species concerned. This can be a particular issue where rare species attract many visitors.

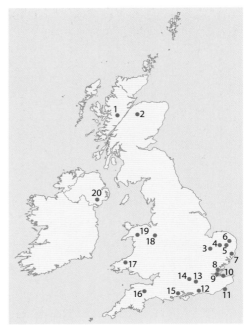

SITE		OS grid reference
1	Coire Loch, Glen Affric NNR, Highland	NH 294282
2	Abernethy Forest RSPB Reserve, Highland	NH 91
3	Wicken Fen NNR/National Dragonfly Centre, Cambridgeshire	TL 563705
4	Thompson Common Norfolk Wildlife Trust Reserve, Norfolk	TL 941966
5	Strumpshaw Fen RSPB Reserve, Norfolk	TG 3306
6	Upton Broad & Marshes Norfolk Wildlife Trust Reserve, Norfolk	TG 380137
7	Staverton Lakes, Suffolk	TM 359515
8	Wat Tyler Country Park, Essex	TQ 7386
9	Cliffe Marshes, Kent	TQ 77
10	Isle of Sheppey, Kent	TQ 96 / 97
11	Dungeness RSPB Reserve, Kent	TR 0619
12	River Arun at New Bridge, West Sussex	TQ 069260
13	Thursley Common NNR, Surrey	SU 9041
14	Bramshill Common & Warren Heath, Hampshire	SU 75/76
15	Crockford Stream, Hampshire	SZ 351989
16	Grand Western Canal Country Park, Devon	ST 0515
17	Preseli Mountains, near Brynberian, Pembrokeshire	SN 1034
18	Fenn's & Whixall Mosses NNR, Shropshire	SJ 489358
19	Llyn Tecwyn Isaf, Gwynedd	SH 629368
20	Lackan Bog, Co. Down	J 240370

SPECIES		Site 1	2	3	4	5	6	7	8	9	10	11	12	13	14	15	16	17	18	19	20
Damselflies	Scarce Emerald Damselfly				●				●	●	●										
	Southern Emerald Damselfly									●											
	Willow Emerald Damselfly							●													
	Small Red Damselfly													●	●	●		●		●	
	White-legged Damselfly												●								
	Southern Damselfly															●		●			
	Northern Damselfly		●																		
	Irish Damselfly																				●
	Dainty Damselfly										●										
	Variable Damselfly			●		●	●														●
	Scarce Blue-tailed Damselfly															●		●			
	Red-eyed Damselfly			●	●	●	●		●	●		●		●	●		●				
	Small Red-eyed Damselfly			●	●	●			●	●	●	●			●						
Dragonflies	Hairy Dragonfly			●	●	●	●	●	●	●	●	●	●				●			●	●
	Azure Hawker	●																			
	Southern Migrant Hawker								●	●											
	Norfolk Hawker					●	●														
	Lesser Emperor										●										
	Common Club-tail												●								
	Downy Emerald	●											●	●	●			●			
	Brilliant Emerald	●											●	●	●						
	Northern Emerald	●	●																		
	Scarce Chaser					●							●				●				
	Keeled Skimmer													●	●	●		●		●	
	White-faced Darter	●	●																●		
	Red-veined Darter										●										

Male **Common Club-tail**

Watching and photographing Dragonflies

Fine weather is usually needed for field visits, especially to see territorial dragonflies, which fly mainly when the sun shines. When it is very hot, however, dragonflies can be too active to be viewed well, and may even retire to shade. Sometimes, the best views can be had outside the main period of activity, which is usually between about 10 a.m. and 3 p.m., or during cloudy and even showery spells on an otherwise hot day. The main season for dragonfly watching is May to August: the 'spring' species are best between late May and late June, with the 'summer' species in July–August.

Visit a wide range of good sites to broaden your knowledge of the scarcer, more localised species. The section on *pages 32–33* suggests some good sites to visit. It is always helpful to meet and learn from more experienced people, so consider joining the British Dragonfly Society (BDS) and attending its meetings (see *page 216*), at which expert leaders and other observers will help you see Dragonflies and learn more about them. The Field Studies Council (FSC) runs field courses on Dragonflies.

As it is usually difficult to get very close views of most Dragonflies, visiting small sites will improve your chances of seeing them at close range. Wherever you are, though, close-focus binoculars will be a great advantage. Binoculars of 8× or 10× magnification are suitable if they focus down to 2 m. Some binoculars and monoculars focus even closer and may be used as a field microscope, but their narrow field of view makes it more difficult to locate Dragonflies, especially when they are in flight. A spotting scope can also be useful for looking at a distant Dragonfly when a close approach is not possible (they can also be used with compact and phone cameras to produce 'digiscope' images).

For a critical examination, and also to appreciate the real beauty of Dragonflies, consider catching a few individuals with an insect net. This is easier with damselflies, which move more slowly, and in any case is best achieved by sweeping the net upwards from behind or below and flicking over the net bag to prevent escape. Extract your catch by holding the wings together – they are surprisingly tough – and view it through a hand-lens, monocular or the 'wrong' end of your binoculars. Although netting to collect specimens is now the domain of a few scientists, nearly all Dragonfly species can be captured without the need for a licence. The few exceptions that are afforded special protection are highlighted in the species accounts. Releasing dragonflies onto a suitable perch may give an opportunity to photograph them; damselflies, however, usually fly away immediately. Good photographs or video sequences, especially of a rarity, can help to verify the identification of most Dragonflies.

If you want to take good photographs (or video) of Dragonflies, your field skills will have to be well refined. Close views will invariably be necessary, so it is best to wear sombre clothing, make a slow approach and avoid sudden movements or casting your shadow on the subject. Watch first from a distance to see if your target shows a pattern in its behaviour, such as returning to a particular perch or adopting a regular flight path. Avoid excessive trampling, especially when Dragonflies are emerging, as they can be hard to see and are easily damaged. Good conduct also includes not disturbing your target unduly and being sensitive to the wishes of other observers.

Many people looking for Dragonflies carry a digital camera of some sort. Zooming in on the images taken can instantly reveal critical features that are otherwise hard to see – this can be a great aid to field identification. It is possible to obtain reasonable images with compact cameras and phone cameras, especially if the subject allows a close approach, although focusing can be very difficult with these cameras. High quality results can be obtained with more expensive equipment, such as an image-stabilised, autofocus macro lens on a high-resolution digital single-lens reflex (DSLR) camera. Alternatively, so-called 'bridge' cameras combine the benefits of relatively low cost, light weight, high magnification and close focus with good depth of focus (depth of field). On the down side, backgrounds are 'noisier' than with DSLR cameras, image resolution is poorer and flight shots are almost impossible.

Many insect photographers focus on the eyes of their subjects, using a wide lens aperture (low f-number) to get a minimal depth of field and out-of-focus background. Unfortunately, such images are often unsuitable for identifying Dragonflies, which are surprisingly three-dimensional. Since it is difficult to get the whole creature in focus, try to photograph it from both directly above and from the side of the subject. Using flash can reduce camera-shake and improve the focus range, but can have the disadvantages of producing unnatural reflections from the wings, dark backgrounds and altered colours. The use of a tripod or monopod should help to reduce camera shake when using high magnifications and/or slow shutter speeds. Many photographers now exhibit their images on websites for others to appreciate and to support claims of rare species. However, beware that many Dragonfly images on websites are incorrectly identified! Perhaps this book will help … .

Banded Demoiselle: the photograph on the left was taken with an inexpensive compact camera in natural light; that on the right shows the effects of using flash on the same individual.

How to identify Dragonflies

An introduction to identification

The key to learning how to identify any group of species is first to gain a good knowledge of the commoner species. Although there are relatively few species to worry about in Britain, this does not mean that there is no scope for confusion – the variations in colour related to age and sex can be confusing for beginner and expert alike!

Remember that tenerals have subdued colours (check for dull eye colour and pale wing-spots (pterostigma) and reflective wings, and a rather weak, unsteady flight). If it is possible to obtain close views, it is often worth sexing individuals: look for the bulge of the male's accessory genitalia under the front of the abdomen, or the female's ovipositor under the rear. Try to judge size only when one individual is alongside another for direct comparison. It is helpful to learn the different species' flight and perching characteristics and their habitat preferences or tolerances (*e.g.* acidic; enriched; brackish; standing or running water) – see the table on *pages 30–31* for details of habitat preferences.

The most useful features to check when identifying Dragonflies are summarised in the following table:

Damselflies		Dragonflies
Eye colour	Head	Eye colour; frons ('face') colour and pattern
Presence and colour of stripes on the top and sides; shape of pronotum	Thorax	Presence, extent and colour of stripes on the top and sides
Colour; markings on top, especially of the second, and eighth to tenth abdominal segments (abbreviated as S2 and S8–10)	Abdomen	Colour; markings on top, especially of the second, and eighth to tenth abdominal segments (abbreviated as S2 and S8–10); presence of a 'waist'
Colour and shape of wing-spots	Wings	Colour at base, and of veins (especially the costa); colour and shape of wing-spots
Colour	Legs	Colour(s)

It can be difficult to remember diagnostic combinations of features. These are summarised in the charts on *pages 44–55*, but many people find them easier to remember if they can be related to something silly. For example, think of Azure Damselfly male as a snooker player and Variable Damselfly male as Dracula! Explanations are given in the relevant species account.

The flight periods given in each species account will sometimes help in identification. For example, a well-patterned hawker flying in May is most likely to be a Hairy Dragonfly, rather than one of the late summer hawkers. However, larval development may be accelerated as a consequence of climate change, resulting in unusually early or late emergence.

Of course, once you have finally mastered the identification of adult Dragonflies, there is always the challenge of identifying larvae and exuviae! (See *pages 190–211*.)

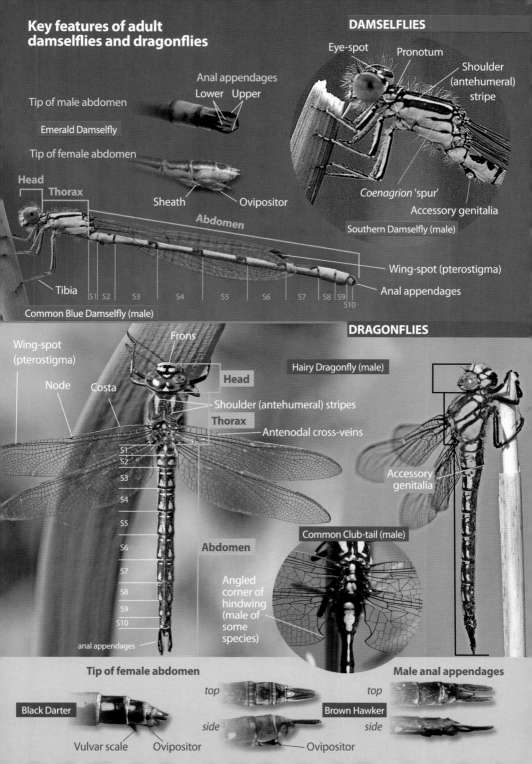

Key features of adult damselflies and dragonflies

DAMSELFLIES

Eye-spot
Pronotum
Shoulder (antehumeral) stripe

Coenagrion 'spur'

Accessory genitalia

Southern Damselfly (male)

Anal appendages
Lower Upper

Tip of male abdomen

Emerald Damselfly

Tip of female abdomen

Head

Thorax

Sheath Ovipositor

Abdomen

Tibia

S1 S2 S3 S4 S5 S6 S7 S8 S9 S10

Common Blue Damselfly (male)

Wing-spot (pterostigma)

Anal appendages

DRAGONFLIES

Frons

Hairy Dragonfly (male)

Wing-spot (pterostigma)

Node Costa

Head

Shoulder (antehumeral) stripes

Thorax

Antenodal cross-veins

S1
S2
S3
S4
S5
S6
S7
S8
S9
S10

anal appendages

Accessory genitalia

Common Club-tail (male)

Abdomen

Angled corner of hindwing (male of some species)

Tip of female abdomen

Black Darter

Vulvar scale Ovipositor

top

side

Ovipositor

Male anal appendages

Brown Hawker

top

side

Colour variations

The changes in colour between emergence and sexual maturation referred to previously (*page 17*) frequently results in confusion and sometimes leads to misidentification. If you cannot identify a Dragonfly, it is important to look for all potential clues, including the colour of the wing-spots or legs, as well as the pattern of black on the sides of the thorax; it is also helpful to confirm the sex of an individual (see *pages 36–37*).

Many colour forms are so well defined that they have been given scientific names. In the case of the blue-tailed damselflies, some of these colour forms merely represent immature phases.

Potentially more confusing, however, are the different colour forms in female red and blue damselflies. In essence, these comprise a typical 'dark form', in which the upper surface of the abdomen is largely black with only traces of red or blue at the segment joints, and one or two forms with extensive areas of colour. In Small Red Damselfly, one of the female forms has such a red abdomen that it resembles the male.

Although they do not have the named colour forms of females, male blue damselflies can show aberrations in their patterning. A small proportion of any population, or sometimes all individuals in isolated populations, have more or less black on S2. This can lead to misidentifications, so it is important to take care when rare species are suspected in areas where they have not been seen before. In such cases, details of the pronotum and/or anal appendages should be checked to confirm identification.

Many female dragonflies become darker as they age. Some female chasers and skimmers may develop pruinescence (see Glossary, *page 58*) and female darters a reddish abdomen. Abrasion from the legs of females during mating results in 'scars' in the pruinescence of male chasers and skimmers. The blue colour of Common Blue Damselfly is affected by cool temperatures, males becoming grey.

Blue-tailed Damselfly female forms: *rufescens* (top) changes to *rufescens-obsoleta* (bottom) when mature

Large Red Damselfly female colour forms: *melanotum* (left), *typica* (centre) and *fulvipes* (right)

Southern Damselfly males: typical 'Mercury' mark on S2 (left) and variation (right)

The types of Dragonfly

This section provides an introduction to the types of Dragonfly that are known to have bred in Britain or Ireland. The 20 species of damselfly fall into nine groups, and the 26 dragonflies into ten groups. A male from each of these groups is shown, with a brief description of the key identification features of the group and a page reference to the relevant species account(s).

Damselflies (Zygoptera)

DEMOISELLES
Pages 60–62

Genus: *Calopteryx*   2 species

Unmistakable large damselflies with metallic green or blue bodies, dark wings (green or bronze in females). The fluttering wing-beats are also distinctive.

Beautiful Demoiselle

EMERALD DAMSELFLIES
Pages 64–70

Genera: *Lestes* and *Chalcolestes*   4 species

Medium-sized, metallic green damselflies that typically rest with their wings held open at about 60° to the abdomen.

Emerald Damselfly

LARGE RED DAMSELFLY
Page 72

Genus: *Pyrrhosoma*   1 species

A medium-sized, red damselfly with black legs and wing-spots and a black line across the side of the thorax.

Large Red Damselfly

SMALL RED DAMSELFLY
Page 74

Genus: *Ceriagrion*   1 species

A small and often inconspicuous dainty red damselfly with reddish legs, eyes and wing-spots, and two black lines across the side of the thorax. Stays low down, below waist-height.

Small Red Damselfly

WHITE-LEGGED DAMSELFLY

Genus: *Platycnemis*

Page 76

1 species

A medium-sized, pale blue or whitish damselfly with flattened tibiae, pale brown wing-spots and an additional pair of 'shoulder stripes'.

White-legged Damselfly

BLUE DAMSELFLIES

Genus: *Coenagrion*

Pages 78–88

6 species

Small to medium-sized damselflies with narrow shoulder stripes, black wing-spots and a distinctive black 'spur' marking on the side of the thorax (see *page 37*).

Azure Damselfly

COMMON BLUE DAMSELFLY

Genus: *Enallagma*

Page 90

1 species

A medium-sized damselfly with broad shoulder stripes, and lacking the distinctive black 'spur' marking on the side of the thorax shown by all other blue damselflies (see *page 37*). Often flies out over open water away from the margins.

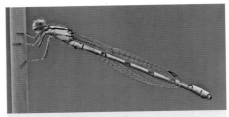

Common Blue Damselfly

BLUE-TAILED DAMSELFLIES

Genus: *Ischnura*

Pages 92–94

2 species

Small to medium-sized damselflies with diagnostic two-toned wing-spots. In forms with blue near the tip of the abdomen, this is confined to S8 or S8 & 9. More likely to be seen mating than other damselflies, especially in late afternoon.

Blue-tailed Damselfly

RED-EYED DAMSELFLIES

Genus: *Erythromma*

Pages 96–98

2 species

Medium-sized damselflies. Males have red eyes, dark top to the thorax and mainly dark abdomen with blue on the top confined to S9–10. Females have a dark abdomen with blue segment divisions towards the tip. Typically sit on floating vegetation.

Red-eyed Damselfly

Dragonflies (Anisoptera)

HAIRY DRAGONFLY

Page 100

Genus: *Brachytron* 1 species

A medium-sized dragonfly with hairy thorax, paired dots on the abdomen, and long, thin, brown wing-spots. Territorial males patrol low down along ditches or the edge of tall emergent vegetation. The flight period is in spring, before the main flight season of most other hawkers. Hangs vertically when perched.

Hairy Dragonfly

HAWKERS

Pages 102–114

Genus: *Aeshna* 7 species

Medium- to large-sized dragonflies with powerful flight. The abdomen is parallel-sided (though sometimes very slightly waisted). In four of the breeding species, the abdomen has pairs of blue, green or yellow dots along its length. In the other two, the abdomen is brown. All species have distinctive and usually prominent markings on the side of the thorax and do not appear hairy. Territorial males are often seen patrolling, generally below waist height. Adults typically feed late in the day high up along woodland edges away from water. Hangs vertically when perched.

Migrant Hawker

EMPEROR DRAGONFLIES

Pages 116–118

Genus: *Anax* 2 species

Large dragonflies with green eyes and green or brownish thorax which is virtually unmarked. The abdomen is parallel-sided, though sometimes slightly waisted, blue or green, or brown with a blue base, but with a distinctive broad, irregular black line along the top. Territorial males are often seen patrolling for long periods above waist height. Hangs vertically when perched.

Emperor Dragonfly

GOLDEN-RINGED DRAGONFLY *Page 120*

Genus: *Cordulegaster*   1 species

The longest British dragonfly, readily identified by its black and yellow markings and green eyes, which touch. Males patrol low, back and forth in territories along flowing waters, landing periodically in bankside vegetation. Individuals perch to eat prey, when they may allow a close approach. Hangs vertically when perched.

Golden-ringed Dragonfly

COMMON CLUB-TAIL   *Page 122*

Genus: *Gomphus*   1 species

A medium-sized dragonfly with yellow and black markings; the only British dragonfly which has eyes that do not touch. It also has complex black and yellow or pale green patterning. Males patrol rapidly up and down territories along rivers. Often found well away, sometimes many kilometres, from the rivers where they breed.

Common Club-tail

EMERALD DRAGONFLIES *Pages 124–128*

Genera: *Cordulia* & *Somatochlora* 3 species

Medium-sized, metallic green dragonflies with green eyes. Usually seen in flight patrolling regular beats along the margins of water bodies or boggy areas, pausing to hover for a few seconds at intervals. They may be seen feeding high up around the canopy of trees, where they land to eat their prey.

Brilliant Emerald

CHASERS
Pages 130–134

Genus: *Libellula* — 3 species

Medium-sized blue or brownish dragonflies with dark markings at the base of the hindwings, although these markings are not always easy to detect. They are regularly seen on prominent perches, from which they sally forth to catch food, chase off intruders or pursue females, often returning to the same perch.

Broad-bodied Chaser

SKIMMERS
Pages 136–138

Genus: *Orthetrum* — 2 species

Small to medium-sized blue (male) or yellowish-brown (female) dragonflies. Males are readily identified by a combination of blue abdomen and lack of dark markings at the base of the hindwings. Females can be confused with female darters, but have distinct yellow antenodal cross-veins. Often seen skimming low over the water, or perched on the ground or among low vegetation.

Keeled Skimmer

WHITE-FACED DARTER
Page 140

Genus: *Leucorrhinia* — 1 species

A small, red-and-black (mature male) or yellow-and-black (female or immature) dragonfly with dark bases to the hindwings and a distinctive white 'face'. It is restricted to peaty pools, perching low down and generally behaving like *Sympetrum* darters.

White-faced Darter

DARTERS
Pages 142–150

Genus: *Sympetrum* — 5 species

Small, red or black-and-yellow (mature male) or yellowish-brown (female or immature) dragonflies, lacking dark bases to the hindwings. They sit typically on bare surfaces or prominent perches, from which they dart to catch prey, often returning to the same perch. Eggs are often laid with the pair 'in-tandem'.

Common Darter

	DEMOISELLES	ABDOMEN
MALES	Beautiful Demoiselle *page 60*	Metallic blue-green
	Banded Demoiselle *page 62*	
FEMALES	Beautiful Demoiselle *page 60*	Metallic green, with bronzy tip and narrow brownish stripe on top of S8–10
	Banded Demoiselle *page 62*	Metallic green, with bronzy tip and narrow pale stripe on top of S8–10

EMERALD DAMSELFLIES:	ABDOMEN (8× life-size)		
MALES	ANAL APPENDAGES / S10 (from above)		S2 / S1
Emerald Damselfly *page 64*		Apps. black; lower pair straight, narrow tips; S10 blue	S2 all powder blue
Scarce Emerald Damselfly *page 66*		Apps. black; lower pair incurved, broad tips; S10 blue	S2 front half powder blue
Southern Emerald Damselfly *page 68*		Apps. pale; lower pair tips point outwards; S10 metallic bronzy-green with pale edges	S2 metallic bronzy-green
Willow Emerald Damselfly *page 70*		Apps. pale; lower pair short, dark; S10 metallic bronzy-green	S2 metallic green
FEMALES	OVIPOSITOR / S9–S10 (side view)		S2 / S1
Emerald Damselfly *page 64*		Ovipositor extends to tip of S10, dark below; sheath pointed	S2 metallic green; S1 spots triangular
Scarce Emerald Damselfly *page 66*		Ovipositor extends beyond S10, dark below; sheath pointed	S2 metallic green; S1 spots rectangular
Southern Emerald Damselfly *page 68*		Ovipositor extends to tip of, or just beyond, S10, pale; sheath rounded	S2 metallic bronzy-green
Willow Emerald Damselfly *page 70*		Ovipositor extends to tip of, or just beyond, S10, dark patches; sheath rounded	S2 metallic green

The coloured blocks relate broadly to the 'types' of damselfly (see *pages 39–40*). The principal differences between the 'types' and species are given; red text indicates diagnostic features.

THORAX	WINGS	WING-SPOT	LEGS	EYES
Metallic blue-green	Wholly brown-black; tip & base sometimes paler		Black	Dark red
	Broad blue-black band			
Metallic green	Wholly brownish	white false wing-spot	Black	Dark red
	Wholly greenish	white false wing-spot		

THORAX		BACK OF HEAD	WING-SPOT	LEGS	EYES
ABOVE	SIDE				
Metallic green above, with powder blue pronotum; powder blue below		Green	narrow: length/width >3	Mainly black	Blue
			broad: length/width <3		
Metallic bronzy-green above, with broad yellowish shoulder stripes		Yellow	black and white	Cream with black stripes	Brownish
Metallic green above, with narrow shoulder stripes and prominent 'spur' marking		Green	large and pale	Black and cream	
Metallic green above; yellow below		Green	narrow: length/width >3	Black and cream	Brown
			broad: length/width <3		
Metallic bronzy-green above, with broad yellowish shoulder stripes		Yellow	black and white	Cream with black stripes	
Metallic green above, with incomplete, narrow shoulder stripes and prominent 'spur' on side		Green	large and pale	Black and cream	

ADULT IDENTIFICATION CHART: Male damselflies

A summary of the key features of mature adult male damselflies breeding in Britain.

SPECIES	ABDOMEN (2× life-size)	S2 / S1 (8× life-size)
Large Red Damselfly *page 72*		
Small Red Damselfly *page 74*		
White-legged Damselfly *page 76*		
Southern Damselfly *page 78*		S2 S1
Northern Damselfly *page 80*		S2 S1
Irish Damselfly *page 82*		S2 S1
Dainty Damselfly *page 84*		S2 S1
Azure Damselfly *page 86*		S2 S1
Variable Damselfly *page 88*		S2 S1
Common Blue Damselfly *page 90*		S2 S1
Blue-tailed Damselfly *page 92*		
Scarce Blue-tailed Damselfly *page 94*		
Red-eyed Damselfly *page 96*		
Small Red-eyed Damselfly *page 98*		

The coloured blocks relate broadly to the 'types' of damselfly (see *pages 39–40*). The principal differences between the 'types' and species are given; red text indicates diagnostic features.

ABOVE	THORAX SIDE	PRONOTUM (Rear edge)	WING-SPOT	LEGS	EYES
Two red shoulder stripes	One black stripe		Black	Black	Red
Wholly bronzy-black	Two black stripes		Red	Red or pink	Red
Four blue stripes			Pale brown	Broad, white tibiae	
Narrow blue shoulder stripes			Black	Black and pale blue; narrow tibiae	Green-blue
Broken blue shoulder stripes					
Broad blue shoulder stripes					
Narrow blue shoulder stripes				Black and pale blue	Green-blue
Wholly black			Black	Black	Deep burgundy-red
Narrow shoulder-stripes, usually incomplete				Dark bluish-grey	Tomato-red

A summary of the key features of mature adult female damselflies breeding in Britain.

SPECIES	ABDOMEN (2× life-size)	
Large Red Damselfly *page 72*	Form *typica*	
	Form *fulvipes*	
	Form *melanotum*	
Small Red Damselfly *page 74*	Form *typica*	
	Form *melanogastrum*	
	Form *erythrogastrum*	
White-legged Damselfly *page 76*	Normal form	
	Form *lactea*	
Southern Damselfly *page 78*		
Northern Damselfly *page 80*		
Irish Damselfly *page 82*		
Dainty Damselfly *page 84*		
Azure Damselfly *page 86*	Dark form	
	Blue form	
Variable Damselfly *page 88*	Blue form	
	Dark form	
Common Blue Damselfly *page 90*	Green form	
	Blue form	
Blue-tailed Damselfly *page 92*	Form *violacea, typica* & *rufescens*	
	Forms *infuscans* & *rufescens-obsoleta*	
Scarce Blue-tailed Damselfly *page 94*	Normal form	
	Form *aurantiaca*	
Red-eyed Damselfly *page 96*		
Small Red-eyed Damselfly *page 98*		

ABOVE	THORAX — SIDE	PRONOTUM (Rear edge)	WING-SPOT	LEGS	EYES
Red or yellow shoulder stripes	One **black** stripe		Black	Black	Dull red
Very narrow shoulder stripes, broken or absent	Two **black** stripes		Red	Pinkish	Brownish-red
Four **pale** stripes			Pale brown	Broad, white tibiae	
Narrow **pale** green or yellow shoulder stripes			Black	Black and pale blue; narrow tibiae	Greenish
Broad **pale** green, buff or blue shoulder stripes					
Shoulder stripes absent in some forms				Black and pale stripes	Brownish-red
Brownish-green (mainly orange in immature)					
Incomplete, narrow yellow shoulder stripes			Pale brown	Black	Brownish-red
Narrow shoulder stripes				Dark bluish-grey	Brownish above; pale below

Hawkers, emperors and Golden-ringed Dragonfly

SPECIES	WINGS	HEAD	THORAX	ABDOMEN (life-size)
MALES				
Hairy Dragonfly *page 100*	COSTA: Yellow. WING-SPOT: Long, thin and brown	EYES: Blue from above	Hairy	
Azure Hawker *page 102*	COSTA: Brown	EYES: Blue from above; contact zone short		
Common Hawker *page 104*	COSTA: Yellow	EYES: Blue from above; contact zone broad		
Migrant Hawker *page 106*	COSTA: Brown	EYES: Blue from above		
Southern Migrant Hawker *page 108*	COSTA: Pale	EYES: Blue		
Southern Hawker *page 110*	COSTA: Brown	EYES: Blue from above		
Brown Hawker *page 112*	Brown-tinted	EYES: Bluish from above	TOP: Wholly brown. SIDE: 2 prominent yellow stripes	
Norfolk Hawker *page 114*	Amber base to hindwings. WING-SPOT: Amber	EYES: Green	TOP: Brown, usually without stripes. SIDE: 2 indistinct yellow stripes	
Emperor Dragonfly *page 116*	COSTA: Yellow. WING-SPOT: Brown	EYES: Blue-green from above	Wholly green with 2 small blue wedges on top	
Lesser Emperor *page 118*	Yellow suffusion COSTA: Yellow. WING-SPOT: Brown	EYES: Green from above	TOP: Violet-brown. SIDE: Green-brown	
Golden-ringed Dragonfly *page 120*	COSTA: Yellow. WING-SPOT: Black	EYES: Green from above, just touching	TOP: 2 tapering yellow stripes SIDE: 2 bold yellow stripes	

The coloured blocks relate broadly to the 'types' of dragonfly (see *pages 41–42*). The principal differences between the 'types' and species are given; red text indicates diagnostic features.

51

FEMALES

ABDOMEN (life-size)	THORAX	HEAD	WINGS
	Hairy	EYES: Brownish from above	COSTA: Yellow WING-SPOT: Long, thin and brown
		EYES: Brown from above; short contact zone	COSTA: Brown
		EYES: Green-brown from above; broad contact zone	COSTA: Yellow
		EYES: Green-brown from above	COSTA: Brown
		EYES: Brown from above	COSTA: Pale
		EYES: Brownish-green from above	COSTA: Brown
	TOP: Wholly brown SIDE: 2 prominent yellow stripes	EYES: Yellowish-brown from above	Brown-tinted
	TOP: Brown, usually without stripes SIDE: 2 indistinct yellow stripes	EYES: Green from above	Amber base to hindwings WING-SPOT: Amber
	Wholly green	EYES: Greenish from above	COSTA: Yellow WING-SPOT: Brown
	TOP: Violet-brown SIDE: Green-brown	EYES: Green from above	Yellow suffusion COSTA: Yellow. WING-SPOT: Brown
	TOP: 2 tapering yellow stripes SIDE: 2 bold yellow stripes	EYES: Green from above, just touching	COSTA: Yellow WING-SPOT: Black

MALES				
SPECIES	**WINGS**	**HEAD**	**THORAX**	**ABDOMEN (life-size)**
Downy Emerald *page 128*	Clear with yellow bases	EYES: Bright green from above 'FACE': Dark, with yellowish jaws and pale bar between eyes	Downy; metallic bronze-green	
Brilliant Emerald *page 126*	Yellow suffusion, deeper at bases COSTA: Yellow	EYES: Bright green from above 'FACE': As Downy Emerald but with additional yellow 'U' on frons	Shiny emerald green	
Northern Emerald *page 128*	Yellow suffusion, deeper at bases	EYES: Bright green from above 'FACE': Yellow spot on each side	Metallic bronze-green	
Common Club-tail *page 122*	COSTA: Dark WING-SPOT: Dark	EYES: Dull green from above, not touching	TOP: Four pale green stripes SIDE: Pale green with thin black line	
Four-spotted Chaser *page 130*	Dark bases to hindwings and spot on nodes	EYES: Brown from above	Brown	
Broad-bodied Chaser *page 132*	Dark bases to wings. WING-SPOT: Black	EYES: Brown from above	Brown with 2 pale stripes on top	
Scarce Chaser *page 134*	Dark bases and spot at wing-tips WING-SPOT: Dark brown	EYES: Blue-grey from above	Brown	
Black-tailed Skimmer *page 136*	Clear. COSTA: Yellow WING-SPOT: Black	EYES: Greenish-blue from above	Olive-brown above with 2 incomplete thin black stripes	
Keeled Skimmer *page 138* (see also *page 55*)	Clear, with line of yellow cross-veins at front of basal half. COSTA: Pale yellowish WING-SPOT: Orange	EYES: Blue-grey from above	Dark brown above with 2 buff stripes, becoming blue with age	

The coloured blocks relate broadly to the 'types' of dragonfly (see *pages 42–43*). The principal differences between the 'types' and species are given; red text indicates diagnostic features.

ABDOMEN (life-size)	THORAX	HEAD	WINGS
	Downy; metallic bronze-green	EYES: Bright green from above 'FACE': Dark, with yellowish jaws and pale bar between eyes	Clear with yellow bases
	Shiny emerald green	EYES: Bright green from above 'FACE': As Downy Emerald but with additional yellow 'U' on frons	Yellow suffusion, deeper at bases COSTA: Yellow
	Metallic bronze-green	EYES: Bright green from above 'FACE': Yellow spot on each side	Clear with faint yellow bases
	TOP: Four yellow stripes SIDE: Yellow with thin black line	EYES: Dull green from above, not touching	COSTA: Dark WING-SPOT: Dark
	Brown		Dark bases to hindwings and spot on nodes. Yellow at base and front of inner wings
	Brown with 2 pale stripes on top	EYES: Brown from above	Dark bases to wings WING-SPOT: Black
	Brown		Dark bases and spot at wing-tips. Yellow suffusion along front of wings WING-SPOT: Brown
	Olive-brown above with 2 incomplete thin black stripes		Clear COSTA: Yellow WING-SPOT: Black
	TOP: Dark brown above with 2 buff stripes SIDE buff, lacking black stripes	EYES: Brown from above	Golden/yellow suffusion to front of basal half with line of yellow cross-veins COSTA: Pale yellow WING-SPOT: Orange

FEMALES

Darters (with female Keeled Skimmer included for comparison)

	MALES				
SPECIES	**WINGS**	**HEAD***	**THORAX**	**LEGS**	**ABDOMEN (1·2× life-size)**
White-faced Darter *page 140*	Dark basal patches; line of yellow cross-veins near front WING-SPOT: Dark brown	FRONS: Cream-white	Black above with 2 broken reddish stripes	Black	
Black Darter *page 142*	Clear WING-SPOT: Black	FRONS: Darkens with age	TOP: Black SIDE: Yellow with black central patch containing three yellow spots	Black	
Common Darter *page 144*	Clear with tiny area of yellow at base WING-SPOT: Yellow to reddish-brown	FSL: absent (except in dark forms in NW Scotland)	TOP: Brown SIDE: 2 large yellow patches on side divided by reddish-brown panel	Dark with yellow stripe along length	
Ruddy Darter *page 146*	Clear with tiny area of yellow at base (may take on golden hue with age) WING-SPOT: Red-brown	FSL: present	TOP: Red-brown with black collar SIDE: Yellowish	Black	
Red-veined Darter *page 148*	Yellow bases and red costa and other veins WING-SPOT: Pale, outlined in black	FSL: present EYES: blue undersides	Reddish above SIDE: Grey-brown	Dark with yellow stripe along length	
Yellow-winged Darter *page 150*	Basal half tangerine WING-SPOT: Red or brown with thick black margins	FSL: present	TOP: Yellow-brown SIDE: Relatively plain yellowish		

* Darter heads: The combination of eye colour, frons colour and the presence of dark lines down each side of the frons helps with darter identification. **The presence or absence of this 'frons side line'** (FSL) **is indicated in the table above.**

Keeled Skimmer
(for comparison

(see also *pages 52–5*

The coloured blocks relate broadly to the 'types' of dragonfly (see *page 43*). The principal differences between the 'types' and species are given; red text indicates diagnostic features.

FEMALES

ABDOMEN (1·2× life-size)	THORAX	LEGS	HEAD*	WINGS
	Black above with 2 broken yellowish stripes	Black	FRONS: Cream-white	Dark basal patches; line of yellow cross-veins near front WING-SPOT: Dark brown
	TOP: Yellow-ochre with black triangle SIDE: Yellow with black central patch containing three yellow spots	Black	FRONS: Yellow, becoming darker with age	Clear with tiny area of yellow at base WING-SPOT: Black
	TOP: Brown SIDE: Yellow divided by narrow black lines	Dark with yellow stripe along length	FSL: absent (except in dark forms in NW Scotland)	Clear with tiny area of yellow at base WING-SPOT: Yellow to reddish-brown
	TOP: Yellow-ochre with black collar SIDE: Yellowish	Black	FSL: present	Clear with tiny area of yellow at base (may take on golden hue with age) WING-SPOT: Red-brown
	Yellow-ochre with dull greenish-yellow sides	Dark with yellow stripe along length	FSL: present EYES: blue undersides	Yellow veins on basal half of wings WING-SPOT: Pale, outlined in black
	TOP: Yellow-brown SIDE: Relatively plain yellowish		FSL: present	Basal half tangerine (fading with age) WING-SPOT: Red or brown with thick black margins
	TOP: Dark brown above with 2 buff stripes SIDE buff, lacking black stripes		EYES: Brown from above	Golden/yellow suffusion to front of basal half with line of yellow cross-veins. COSTA: Pale yellow WING-SPOT: Orange

Identifying male emerald dragonflies, hawkers and emperors in flight

Three groups of dragonflies – the hawkers, the emperors and the emeralds – are more often seen in flight than perched, particularly males when patrolling territories. This makes them difficult to identify, although it is possible to see some of the characteristic features when viewed at close range or with binoculars. All species are shown here at life-size.

Southern Migrant Hawker
Eyes all blue; sides of thorax mainly bluish; abdomen held rather straight

Migrant Hawker
Broad pale stripes on sides of thorax and faint markings on top; abdomen uptilted and slightly droop-tipped

HAWKERS

Focus on the thorax and abdomen patterning and the angle at which the abdomen is held. On the thorax, check for the presence or thickness of stripes on the top and sides. On the abdomen, look at the extent of blue or yellow markings near the tip (See pages 100–114**)**

Common Hawker
Narrow pale stripes on top and sides of thorax; abdomen uptilted and straight; yellow costa may be visible

Southern Hawker
Very broad pale 'head-light' stripes on top and very pale sides to thorax; abdomen more or less horizontal and slightly curved, with diagnostic 'tail-lights' (unbroken bands across S9 & 10)

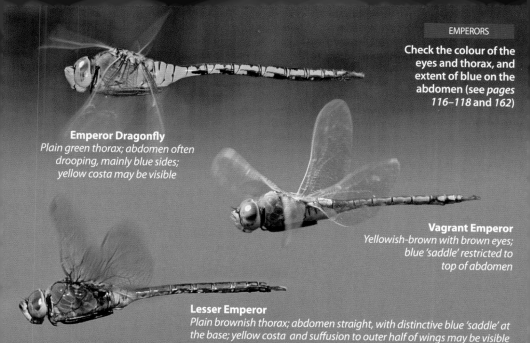

Check the colour of the eyes and thorax, and extent of blue on the abdomen (see *pages 116–118* and *162*)

Emperor Dragonfly
Plain green thorax; abdomen often drooping, mainly blue sides; yellow costa may be visible

Vagrant Emperor
Yellowish-brown with brown eyes; blue 'saddle' restricted to top of abdomen

Lesser Emperor
Plain brownish thorax; abdomen straight, with distinctive blue 'saddle' at the base; yellow costa and suffusion to outer half of wings may be visible

In emeralds, look for the angle at which the abdomen is held, its shape, colour and length and the colour of the thorax (see *pages 124–128*)

Brilliant Emerald
Thorax and slightly waisted abdomen uniformly metallic emerald-green; yellow suffusion to wings may be visible

Downy Emerald
Bronze-green thorax; abdomen bronze, elevated and waisted with distinct bulbous tip

Northern Emerald
Bronze-green thorax; abdomen dark and narrowly-waisted, with distinctive calliper-shaped anal appendages

Glossary

Accessory genitalia The structure on the underside of **S2** and **S3** of the male's abdomen that holds sperm prior to copulation.

Anal appendages The 'claspers' at the tip of the male's abdomen that grasp the female during copulation.

Antenodal cross-veins Veins at right angles to the **costa**, between the body and the node.

Caudal lamella Appendage at the rear of damselfly **larvae** (usually three), used for respiration and locomotion (plural: lamellae).

Cerci The upper pair of **anal appendages** on a dragonfly **larva**.

Costa The front vein, which forms the leading edge of the wing.

Epiproct The central **anal appendage** of a dragonfly **larva**.

Exuvia The cast larval skin (plural: exuviae).

Frons The uppermost part of the front of the head ('face'), most prominent in dragonflies.

Immature An adult that has not yet attained the full colouring typically associated with sexual maturity.

Jizz The often indefinable characteristic impression given by an animal or plant, usually defined by shape or movement.

Labium The extendable lower lip of the **larval** mouthparts, sometimes termed the 'mask', used for catching prey.

Larva (or nymph) The immature stage before becoming an adult (plural: larvae).

Metamorphosis The transition from **larva** to adult; this is termed 'incomplete metamorphosis' in Dragonflies, since there is no pupal stage.

Node (or nodus) The notch half-way along the leading edge of the wing.

Obelisk position Position adopted by some dragonflies where the abdomen is pointed towards the sun to avoid overheating.

Ovipositor The blade-like egg-laying structure under **S8–10** of female Dragonflies that deposits eggs into plant material or substrates.

Paraprocts The lower pair of **anal appendages** on a dragonfly **larva**.

Prementum The stalked base of the **labium**.

Pronotum A shield-like plate that covers the top of the front of the thorax in dragonflies or prothorax in damselflies.

Pruinescence The waxy bloom that gives some Dragonflies a powder blue colour.

S2, S3, S8–10 etc. Abbreviations used for the number, or range, of the abdominal or antennal segment(s), counted from the base.

Shoulder stripes The pair of stripes on the top of the thorax, technically known as antehumeral stripes.

Teneral A newly-emerged adult Dragonfly, with soft cuticle, lacking the full adult colouration and often with shiny wings.

Vulvar scale The sometimes prominent flap below **S8**, present in dragonflies (never damselflies) that lay eggs freely onto the water surface.

Wing-spot The dark or coloured cell along the leading edge of the wing, towards the tip, technically known as the pterostigma.

The species accounts

The species accounts that follow are grouped as follows:

Pages 60–150: the 46 species that are known to have bred in Britain or Ireland.

Pages 154–186: the two former breeders (which are on pages shaded red), eight confirmed vagrants and seven potential vagrants (the last of which are on pages with a shaded grey background).

The order of species is broadly taxonomic, although in some cases similar species have been grouped together for ease of reference. The text for each is presented as follows:

SCALE and SCALE BAR
The scale of photos on the plate is given next to English name (LS = life-size). Bar shows actual maximum and minimum total length.

CONSERVATION STATUS and LEGISLATION
For details see *pages 212–215*.

DISTRIBUTION MAP
■ = main breeding range.
■ = secondary breeding range and other records.
● = location of isolated record.
● = former location for extinct species.

MEASUREMENTS
Typical adult biometrics.

ILLUSTRATIONS
Highlighting key features.

THE PLATES
Typically showing both sexes and immature forms from above and the side, scaled optimally for each group.

The following annotations are used on the plates:

M male   **F** female

i immature

t teneral   **o** old

Other annotations specific to the plate are explained in the text.

NOMENCLATURE Follows BDS, or Dijkstra & Lewington for species not recorded in Britain or Ireland.

— ×2 **English name**   *Scientific name*

Former name (in brackets), or English names used in Europe (Eu) and/or Ireland (Ir)

Red List (CATEGORY)
Legislative protection
BAP listing

Status in Britain & Ireland

Principal habitat(s) and secondary habitat(s)

J F M A M J J A S O N D

| Overall length: | xx–xx mm |
| Hindwing: | xx–xx mm |

M   **F**

THORAX   THORAX

LOOK-ALIKES
The species or forms most likely to cause confusion.

A summary of the species' status in Britain and Ireland.

Adult Identification:
A concise description of the adult forms, detailing the key identification features, with the most important highlighted in red text.

The key identification features to look for are highlighted on the photographic plate.

Behaviour: Aspects of behaviour that are a clue to identification.

Breeding habitat: A brief summary of the species' habitat preferences.

Population and conservation: Notes on status and distribution, and a résumé of threats and conservation action.

FLIGHT PERIOD

WHERE TO LOOK
Suggestions of typical sites, or key areas to look for scarce species.

OBSERVATION TIPS
Suggestions and tips to help you find and watch the species.

Beautiful Demoiselle *Calopteryx virgo*

Ir: Beautiful Jewelwing

Locally common

River, stream

J F M A M J J A S O N D

Overall length:	45–49 mm
Hindwing:	29–36 mm

Position of 'false wing-spot' in females

A common large and conspicuous damselfly that is typical of fast-flowing waters in southern and western Britain. Living up to its common name, it presents a dazzling spectacle that greatly enhances the rivers and streams that it inhabits.

Adult Identification: MALE: Dark brown-black wings with iridescent blue veins; the extreme tips and bases may be paler and immatures have browner wings. The body is metallic blue-green. **FEMALE:** Dark brown iridescent wings with a white 'false wing-spot' that lies further from the tip than in Banded Demoiselle (*page 62*). The body is metallic green with a bronze tip to the abdomen.

Behaviour: Males are territorial, perching on bankside vegetation and trees. They habitually flick open their wings when perched, and dash off to chase passing insects, often returning to the same perch. Females live away from water unless egg-laying or seeking a mate, and individuals of both sexes frequently stray well away from water. Females lay up to 300 eggs at a time into emergent or floating vegetation, such as water-crowfoot, often submerging to do so.

Breeding habitat: This is one of the few British species that is restricted to running water, being found only along streams and rivers, often acidic, with a sand or gravel bottom. Although mostly found along heathland and moorland streams, it also occurs widely in farmland and woodland, including well-shaded watercourses.

Population and conservation: Often widespread and abundant along suitable watercourses in the south and west, but absent or at least very local over much of the rest of England, Scotland and the northern half of Ireland. Although it is said to be restricted to unpolluted waters, many stretches that have suffered from localized farm pollution incidents appear to recover and remain occupied. Recent, limited range expansion may be due to improved water quality.

WHERE TO LOOK	**OBSERVATION TIPS**
Found along many watercourses in south-west England, Wales and southern Ireland. Elsewhere, it is localized and usually only found along heathland streams.	Sits on sunny perches over fast-flowing water and flits slowly over the water with a distinctive fluttering flight. They can be wary, so a slow approach is needed to obtain close views.

LOOK-ALIKES

Banded Demoiselle (*page 62*)

Mature males have iridescent, wholly dark wings

Females have brownish wings, each with a white spot

M

F

Mi

Fi

Banded Demoiselle *Calopteryx splendens*

Ir: Banded Jewelwing

Common and widespread in the lowlands

River, stream, canal

J F M A M J J A S O N D

| Overall length: | 45–48 mm |
| Hindwing: | 27–35 mm |

Position of 'false wing-spot' in females

A stunning damselfly, often encountered along slow-flowing streams and rivers in the lowlands.

Adult Identification: MALE: Translucent wings each with a broad, dark, iridescent blue-black spot (or band) across the outer part; on immatures, this marking is dark brown. The body is metallic blue-green. **FEMALE:** Translucent pale green iridescent wings with a white 'false wing-spot' nearer the tip than in Beautiful Demoiselle (*page 60*), and a metallic green body; there is a narrow stripe down the middle of the bronze-coloured S8–10 that is usually paler than in Beautiful Demoiselle.

Behaviour: Males are territorial, but large numbers can be found among lush bankside vegetation and on floating plants. They court females by flicking their wings open and performing an aerial dance in front of them. Like many damselflies, females stay away from water unless looking for a mate or egg-laying. The adults often make use of luxuriant vegetation like nettle beds and tall emergent grasses. Females can lay about 10 eggs per minute for up to 45 minutes, laying into a wide variety of emergent or floating plants, often submerging to do so.

Breeding habitat: Along with the more locally distributed White-legged Damselfly, this species is characteristic of mature, slow-flowing streams and rivers, and sometimes canals, with muddy sediment; it can occur with Beautiful Demoiselle if patches of sand or gravel are also present. Individuals of both sexes frequently stray well away from water and may be found at ponds where breeding is unlikely.

Population and conservation: The Banded Demoiselle is abundant along many of its occupied watercourses. It is absent from Scotland and rare in northern England, where there has been some northward range expansion in recent years. It appears to be tolerant of the high nutrient levels typical of the lower stretches of many rivers.

WHERE TO LOOK
Found along slow-flowing watercourses in the lowlands of Ireland, Wales and England. Look for it sitting in sunny patches on sheltered bankside vegetation.

OBSERVATION TIPS
The large size and fluttering wing-beats help to identify both sexes. The greenish females can be inconspicuous against lush vegetation.

LOOK-ALIKES
Beautiful Demoiselle (*page 60*)

Males have a conspicuous dark 'band' on each wing

Females have greenish wings, each with a white spot

Emerald Damselfly

Lestes sponsa

Widespread and fairly common

Lake, **pond**, canal, ditch, **bog**

J F M A M J J A S O N D

Overall length:	35–39 mm
Hindwing:	19–24 mm

An attractive species that breeds in shallow, well-vegetated standing waters in both the uplands and the lowlands. Numbers peak in late summer, later than most other damselflies.

Adult Identification: All emerald damselflies have largely metallic-green bodies, like the demoiselles, but are distinctly smaller, have clear wings and rest habitually with wings well-spread. The narrow wing-spot (length/width >3) is initially pale and develops through brown to blackish. **MALE:** Powder blue pruinescence develops on S1–2, S9–10 (sometimes S8) and on the pronotum and sides of the thorax. The eyes are blue. The anal appendages are prominent, with the smaller, lower pair being straight. **FEMALE:** Distinctly thicker abdomen than male, lacking any blue and with beige underparts (although rarely pruinesces with age). The eyes are brown. The ovipositor is prominent and just reaches the tip of S10. There is a diagnostic pair of rounded dark spots on S1. Other minor features include pale lower lobes to the pronotum, an isolated metallic spot just above the base of the middle leg, and thin, yellowish shoulder stripes. **IMMATURE:** both sexes are metallic emerald with warm-buff underparts and mature over a period of two weeks or so, a much longer period than in Scarce Emerald Damselfly (*page 66*).

Behaviour: The elongated eggs are inserted into the stems of emergent plants (*e.g.* rushes), the female starting above water and often submerging completely.

Breeding habitat: Tends to favour smaller, shallow standing waters with luxuriant vegetation, such as tall grass, rushes and sedges. Sites range from acidic bog pools, through enriched ponds to brackish ditches. Even sites that dry out in late summer can remain occupied, due to the species' late emergence and the fact that it over-winters as eggs.

Population and conservation: Although not threatened, may be affected locally by large-scale dredging of well-vegetated ponds and ditches.

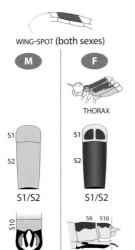

WING-SPOT (both sexes)

M   F

THORAX

S1   S1

S2   S2

S1/S2   S1/S2

S10   S9 S10

ANAL APPENDAGES   OVIPOSITOR

LOOK-ALIKES

Scarce Emerald Damselfly (*p. 66*)

WHERE TO LOOK
Rushy pools almost anywhere in Britain and Ireland during late summer.

OBSERVATION TIPS
Flies very slowly and can be remarkably inconspicuous. Look closely for adults sitting with their wings spread among dense rushes.

Usually holds
the wings
well-spread

Fi

F

Females have
a diagnostic pair
of rounded dark
spots on S1 that
make the front edge
appear semi-circular

Mi

M

M

Males have straight lower
anal appendages (see *opposite*)

Scarce Emerald Damselfly   *Lestes dryas*

Eu: Robust Spreadwing
Ir: Turlough Spreadwing

GB and Irish Red Lists
(Near Threatened)

Rare and local

Lake, pond, ditch

J F M A M J J A S O N D

Overall length:	34–37 mm
Hindwing:	20–27 mm

WING-SPOT (both sexes)

M   **F**

THORAX

S1   S1

S2   S2

S1/S2   S1/S2

S10   S9   S10

ANAL
APPENDAGES   OVIPOSITOR

LOOK-ALIKES

Emerald Damselfly (*page 64*)

A rare and enigmatic species of ephemeral and transitional sites where few other Dragonflies can survive.

Adult Identification: Both sexes have largely metallic green bodies with a bronze iridescence, and habitually rest with wings well-spread. This species is more robustly built than Emerald Damselfly (*page 64*) and has slightly broader wing-spots, but differences in genitalia provide the only reliable means of identification. **MALE:** Powder blue pruinescence develops on S1, usually only the front half or so of S2, S9–10 (sometimes S8), the pronotum and the sides of the thorax. Eyes bluer than all but the oldest Emerald Damselflies. The lower anal appendages are distinctively thick and incurved at the tip. **FEMALE:** Chunky thorax and abdomen, lacking any blue and with greenish-cream underparts. Eyes brown. The ovipositor is prominent and just protrudes beyond the tip of S10. There is a diagnostic pair of square, dark marks on S1. Other minor features include metallic green lower lobes to the pronotum, extensive dark metallic area above the base of the middle leg and very thin incomplete shoulder stripes. **IMMATURE:** both sexes are metallic chocolate-emerald with greenish-cream underparts and lack the 'juvenile' pinkish-buff colour phase of Emerald Damselfly.

Behaviour: Elongated eggs are laid, usually above water, into plant stems, *e.g.* rushes, which are covered as water levels rise in winter.

Breeding habitat: Around the Thames Estuary, it occurs in coastal ditches choked with vegetation. Also found at a few shallow ponds and lakes inland in Norfolk and more widely in Ireland. These sites may dry out occasionally but remain occupied because the species emerges late, overwinters as eggs and has fast-growing larvae.

Population and conservation: Believed extinct in England until re-discovered in Norfolk in 1983. Subsequently found at several sites in south-east England, many of which are at risk from abstraction, eutrophication, over-grazing, successional change, drought and coastal flooding. The species has been found to be more widespread in Ireland than previously thought, although populations vary widely between years.

WHERE TO LOOK	**OBSERVATION TIPS**
Densely vegetated coastal ditches (*e.g.* at Canvey Island, Essex), Breckland ponds (*e.g.* Thompson Common, Norfolk) and turloughs in The Burren, Co. Clare.	Very similar in character to Emerald Damselfly, flying slowly amongst dense emergent vegetation. Close study is essential to confirm identification.

Females look 'chunky'.
They are metallic chocolate-emerald above with greenish-cream underparts and diagnostic square marks on S1

F

F

Females have a long ovipositor

Males have bright blue eyes and the blue on S2 is confined to the front half

Usually holds the wings well-spread

M

M

Males have diagnostic incurved lower anal appendages (see *opposite*)

Southern Emerald Damselfly *Lestes barbarus*

Eu: Migrant Spreadwing

Recent colonist

Pond, ditch

J F M A M J J A S O N D

Overall length:	40–45 mm
Hindwing:	20–27 mm

WING-SPOT (both sexes)

THORAX (both sexes)

S1/S2 (both sexes)

M   F

S10   S9 S10

ANAL   OIVIPOSITOR
APPENDAGES

A large emerald damselfly that occurs just across the English Channel and has made a faltering attempt to colonise Britain.

Adult Identification: Differs from other emerald damselflies in Britain in having bicoloured wing-spots. These are similar to those of the blue-tailed damselflies (*pages 80–83*) but longer (about 4× the width) and much more conspicuous. The wings are typically held well-spread. Both sexes are mainly metallic green, becoming bronze with age. The eyes are greenish, the rear of head is yellow and the edges to the top of S8–10 are pale. The shoulder stripes are relatively broad and yellow. **Male:** generally lacks blue pruinosity, although S10 is whitish. From above, the upper anal appendages are pale with dark tips and the lower anal appendages are small, pointed and out-curved at the tips. **Female:** Ovipositor is entirely pale. **Immature:** Wing-spots wholly pale initially, as with other emerald damselflies.

Behaviour: Breeding biology is similar to other emerald damselflies, but the eggs may be laid into woody stems, such as those of willows, as well as rushes, sedges and other wetland plants.

Breeding habitat: Breeding attempts in Britain have been in dune ponds and ditches. Elsewhere, it is a specialist of seasonally flooded still waters, sometimes slightly brackish. These habitats typically dry out early in the summer and eggs are often laid in apparently dry areas. It is often found some distance from water.

Population and conservation: First seen in Britain at Winterton Dunes, Norfolk, in 2002, and again in 2004, 2009, 2010, 2012 and 2013; however, egg-laying in the last year has been the only evidence of breeding. Egg-laying was also seen at Sandwich Bay, Kent, in 2004, but none has been seen there since seawater flooding the following winter. After singles at Keynsham, Somerset, and Norfolk in 2006, and Norfolk and Suffolk in 2009, egg-laying was seen at Cliffe, Kent, in 2010, and there have annual sightings up to 2013. Records also came from two other sites in Kent in 2010, a site in south Essex in 2010 and 2011, and one was in Sussex in 2011. There have been sporadic records from the Channel Islands and there is evidence that it has been breeding there since 1995.

WHERE TO LOOK	**OBSERVATION TIPS**
Most likely to occur at waters near the south or south-east coasts, at or near temporary pools in sand dunes or coastal marshes.	Similar to the commoner emerald damselflies in holding their wings spread at rest, but appear larger and paler and have diagnostic two-toned wing-spots.

*Usually holds
the wings
well-spread*

M

M

*Both sexes have
diagnostic
dark-and-light
wing-spots*

F

Willow Emerald Damselfly

Chalcolestes viridis

Eu: Western Willow Spreadwing

Recent colonist

Lake, pond, river, stream, canal, **ditch**

J F M A M J J A S O N D

Overall length:	39–48 mm
Hindwing:	23–28 mm

WING-SPOT (both sexes)

THORAX (both sexes)

S1

S2

S1/S2 (both sexes)

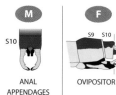

M   F

S10   S9 S10

ANAL APPENDAGES   OVIPOSITOR

LOOK-ALIKES

Emerald Damselfly (*page 64*)
Scarce Emerald Damselfly (*p. 66*)
Southern Em'ld Damselfly (*p. 68*)

Easily overlooked, due to an association with waterside trees, where the large, long-bodied adults may be seen hanging cryptically with their wings spread wide; has recently become well-established in south-east England, centred around Suffolk.

Adult Identification: Both sexes are mainly metallic green, like other emerald damselflies, but larger, darker and without blue eyes or pruinescence. The wing-spots are large and pale, outlined in black (beware of tenerals of other species of emerald damselfly, which also have pale wing-spots and lack pruinescence). The thorax has thin yellow shoulder stripes; the lower border of the metallic green on the side of the thorax is irregular and forms a dark 'spur' (this may be present in a reduced form in the commoner emerald damselflies). **MALE:** Abdomen very long. The upper anal appendages are distinctly pale with black tips and are more than 3× the length of the lower anal appendages. **FEMALE:** The ovipositor is dark with a pale area on the lower edge.

Behaviour: Eggs are laid into twigs overhanging water using serrations on the underside of the ovipositor; this results in distinctive oval galls (galls).

Breeding habitat: Standing and slow-flowing waters with overhanging trees, where males defend vertical territories. In Suffolk, egg-laying has been observed on willows, Alder, Ash, birches, Hawthorn and Elder.

Population and conservation: The first records were of an adult found dead near Pevensey, East Sussex, in 1979; an exuvia collected at Cliffe Marshes, north Kent, in 1992; and a female found on the Suffolk coast in 2007. In 2009, however, some 400 were reported, including tenerals, from 35 sites in east Suffolk, north Essex and south Norfolk, perhaps from an undetected influx in 2007. Egg-laying was seen and the species has since become firmly established, spreading to north Kent and West Suffolk by 2012. In continental Europe, its range extends from north-west Africa to the English Channel and North Sea coasts. It is widespread on Jersey and was first recorded on Alderney in 2010.

WHERE TO LOOK	OBSERVATION TIPS
Scan trees beside or near water, from eye-level to several metres up, for adults hanging on twigs at an angle of about 45 degrees, often in the shade.	Breeding is indicated by series of egg-laying scars on branches. Adults fly in May–November on the continent, but mainly in August–September in England.

galls

M

Both sexes lack
pruinescence
and have a long
abdomen and large,
pale wing-spots

Male anal appendages
are pale with dark tips

F

Usually holds
the wings
well-spread

Large Red Damselfly *Pyrrhosoma nymphula*

Eu: Large Red Damsel
Ir: Spring Redtail

Common and widespread

Lake, **pond**, river, stream, canal, **ditch**, **bog**

J F M A M J J A S O N D

Overall length:	33–36 mm
Hindwing:	19–24 mm

form
fulvipes

form   form
typica   *melanotum*

ABDOMEN (life-size)

LOOK-ALIKES

Small Red Damselfly (*page 74*)
Red-eyed Damselfly (*page 96*)

One of the three most widespread Dragonflies, this is the first species to emerge in spring and brings a welcome dash of colour to otherwise sombre waters at that time of year.

Adult Identification: Relatively large, robust and active, differing from the tiny, weak Small Red Damselfly (*page 74*) in having black, rather than reddish, legs and wing-spot, and broad red or yellow shoulder stripes. Both sexes have red eyes, duller in females and immatures, and yellow sides to the thorax. **MALE:** Deep-red abdomen with bronze-black bands on S7–9, although this can be difficult to see in strong sunlight. The thorax has a bronze-black top and stripe across the side. **FEMALE:** Has three colour forms. The commonest, *typica*, is like the male but has more extensive black on all abdominal segments. The dark form *melanotum* has yellow shoulder stripes (as do immatures of both sexes) and very restricted areas of red and yellow on the largely black abdomen; the red joints between the terminal segments help to distinguish it from female Red-eyed Damselfly (*page 96*), in which they are blue. The red form *fulvipes* has black on the abdomen intermediate between *typica* and the male.

Behaviour: Males emerge slightly earlier and in larger numbers than females, maturing in about 12 days, four days less than females. Males are aggressive, investigating any passing insects and driving off other males. The elongated eggs are laid 'in tandem' into submerged vegetation in batches of about 350.

Breeding habitat: Occupies most wetland habitats, from acidic moorland bog pools to brackish ditches, but avoids fast-flowing waters and shows some preference for sheltered waters with abundant aquatic plants.

Population and conservation: This species' broad distribution reflects a wide habitat tolerance. It is found in virtually every part of Britain and Ireland, although the paucity of records in some arable and upland areas indicates that it is not universally common.

WHERE TO LOOK

This species can be found almost anywhere from spring to mid-summer. Indeed, it not infrequently finds humans – having a habit of landing on pale clothing!

OBSERVATION TIPS

On hot days, territorial males can fly rapidly enough to give the impression of darters, but the peak flight periods do not overlap.

The male's shiny black abdominal markings are sometimes difficult to see, but the black legs rule out the much rarer Small Red Damselfly

M

F typica

F typica

F fulvipes

F melanotum

Small Red Damselfly *Ceriagrion tenellum*

Eu: Small Red Damsel
Ir: Small Redtail

Nationally Scarce B

Scarce and local

Pond, stream, ditch, bog, flush

J F M A M J J A S O N D

Overall length:	25–35 mm
Hindwing:	15–21 mm

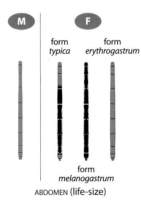

M F

form
typica

form
erythrogastrum

form
melanogastrum

ABDOMEN (life-size)

LOOK-ALIKES

Large Red Damselfly (*page 72*)
Female Southern Dam'lfly (*p. 78*)

This is one of Britain's two red damselflies and amongst the smallest. It flies weakly at heathland bogs and streams in southern England and west Wales.

Adult Identification: A very small damselfly, about 15% shorter than Large Red Damselfly (*page 72*). Both sexes have a diagnostic combination of red or reddish legs, eyes and wing-spot. The thorax is bronze-black on top, with no more than a hint of pale shoulder stripes, and yellow on the sides with two black lines. **MALE:** Abdomen entirely bright red. **FEMALE:** Has three colour forms. The abdomen of the 'normal' form *typica* is bronze-black with S1–3 and S9–10 mostly red. The dark form *melanogastrum* occurs quite frequently and has an almost entirely dark abdomen marked with pale segment divisions, the last three of which are reddish. An uncommon red form, *erythrogastrum*, resembles the male.

Behaviour: Adults fly low and weakly, rarely moving far from breeding waters. Like the Southern Damselfly (*page 78*), which sometimes occurs at the same site, they are reluctant to fly in any but the warmest, sunniest and calmest conditions. They can be surprisingly inconspicuous, especially the darker females. Elongated eggs are laid while 'in tandem' into submerged and emergent plants.

Breeding habitat: Prefers shallow, relatively warm acidic waters, occurring mainly at pools, seepages and small streams. These are usually associated with heathland bogs or old tin or clay workings, up to an altitude of about 300 m. Such sites typically have a lush growth of bog-mosses and Marsh St John's-wort. It also occupies pools at a few calcareous mires.

Population and conservation: This species was found in 78 10 km × 10 km squares during 2000–12 and therefore qualifies as Nationally Scarce B (see *page 213*). It can be abundant at breeding sites but these are often quite isolated. The species is therefore vulnerable to local extinctions, losses having occurred at some sites. It can tolerate short-term scrub encroachment around ponds and has been known to colonise new small ponds from adjacent sites.

WHERE TO LOOK	**OBSERVATION TIPS**
Boggy heathlands of Surrey/ Hampshire, the New Forest, Purbeck and Pembrokeshire and in old tin and clay workings around Bodmin Moor and Dartmoor.	Flies low down and weakly between grass or heathers, where females especially can be very inconspicuous.

Remember: the male "Small Red is all red"
– red abdomen, legs, eyes and wing-spots

Mi

M

F   erythrogastrum

F   typica

Female melanogastrum is confusing
– look for the pinkish legs, wing-spots
and segment divisions at the tip
of the abdomen

F   melanogastrum

F   typica

Eu: Blue Featherleg

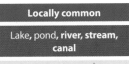

Locally common

Lake, pond, **river, stream, canal**

J F M A M J J A S O N D

Overall length:	35–37 mm
Hindwing:	19–23 mm

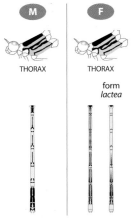

THORAX   THORAX

form *lactea*

ABDOMEN (life-size)

LOOK-ALIKES

Azure Damselfly (*page 86*)
Common Blue Damselfly (*p. 90*)

A species of muddy, slow-flowing waters, beside which the adults may be found in tall, dense vegetation, such as nettles. It is much more restricted in range than the Banded Demoiselle, with which it invariably occurs.

Adult Identification: Mature adults differ from other blue damselflies in having expanded white edges to the tibiae (most obvious on the hind legs of males); paired black markings down most of the abdomen; broad, pale shoulder stripes with additional narrower stripes below; and pale chestnut wing-spots. **MALE:** Black markings on blue, which is often pale and sometimes greenish on the thorax. **FEMALE:** Very pale yellow-green with black markings. **IMMATURE:** Creamy white with brownish thorax and eyes and reduced black abdominal markings; immature females, sometimes referred to as the form *lactea*, have these markings reduced to pairs of spots on S2–6.

Behaviour: After emerging, adults tend to congregate in the shelter of tall vegetation, although some immatures wander away from water and have been found up to 5 km from the nearest breeding sites. Mating is preceded by the male displaying its white legs in response to a specific, jerky flight by the female. Eggs are laid, while 'in tandem', into emergent stems and floating leaves.

Breeding habitat: Favours unshaded, slow-flowing sections of muddy rivers with abundant emergent and floating vegetation. Emergence has been recorded from tidal rivers and the larvae appear to be able to tolerate brackish conditions. Also occurs along muddy streams and some canals, but only rarely in lakes and ponds, although it has increased recently at standing waters.

Population and conservation: Locally abundant along some rivers south of The Wash, but sensitive to the loss of waterside vegetation through intensive grazing or over-zealous cutting. The larvae are susceptible to pollution and the species has been lost from some sites. However, its dispersive behaviour, the canal network and improvements in water quality have enabled it to re-colonise once-polluted stretches and spread to new river catchments in recent years.

WHERE TO LOOK
Adults are most readily found in tall riverside vegetation along mature stretches of rivers such as the Arun, Lea, Severn and Tamar, but they often wander far from water.

OBSERVATION TIPS
Sometimes found in large numbers in sheltered spots. Groups of pairs egg-laying 'in tandem' can often be seen on floating beds of water-crowfoot.

The black 'spur' on the side of the thorax is longer than in Coenagrion species, forming a 'hockey stick' shape in males and most females

The wing-spots are distinctly pale chestnut in both sexes, even when immature

Males have expanded, mainly white tibiae, unique in British species

Southern Damselfly

Coenagrion mercuriale

Eu: Mercury Bluet

GB Red List (ENDANGERED)

PROTECTED: EU legislation

W&C Act 1981 (Sched. 5)

NERC Act 2006 (S41 & S42)

UK BAP Priority Species

Rare and local

Stream, ditch, flush

J F M A M J J A S O N D

Overall length:	29–31 mm
Hindwing:	15–20 mm

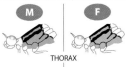

THORAX

PRONOTUM

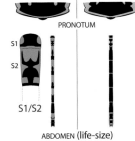

S1

S2

S1/S2

ABDOMEN (life-size)

LOOK-ALIKES

Azure Damselfly (*page 86*)
Form *melanogastrum* of Small
Red Damselfly (*page 74*)

The smallest blue damselfly, with a very restricted distribution on small streams and flushes in southern England, south Wales and Anglesey.

Adult Identification: Both sexes show the '*Coenagrion* spur' (*page 37*) and have a pale bar between rounded eye-spots (absent in Azure Damselfly (*page 86*)) and a small lobe in the centre of the rear edge of the pronotum. **MALE:** Best identified by a 'Mercury' mark on S2, although the shape can vary, with extremes having the 'prongs' narrow and detached (see *page 38*). The other abdominal segments show more black than in Azure Damselfly, with spear-shaped markings on S3 and S4; S9 is mostly black, like Variable Damselfly (*page 88*). The upper anal appendages are longer than the lower. **FEMALE:** Occurs in a blue form, but is usually dull green with the abdomen mainly black above, and with a thistle-like shape on S2 and blue divisions between the last few segments. The *melanogastrum* form of Small Red Damselfly (*page 74*), which often occurs at the same sites, has red segment divisions.

Behaviour: A rather sedentary species, rarely moving more than 50 m when mature, although a few disperse up to 2 km. The flight is weak and low. Females lay alone, inserting their elongated eggs into soft-stemmed submerged herbs.

Breeding habitat: Occurs in very specialised habitats. At heathland sites, most colonies are along runnels and flushes with Bog Pondweed, mostly below 90 m altitude. Others occur on chalk streams and ditches, and in poor fens. Favoured watercourses are base-rich, open, shallow and narrow, with shallow peat or silt over gravel and slow to moderate flows. Permanent flows and stable water temperatures are important, as is nearby shelter for adults.

Population and conservation: In 2005, it was present in 85 1 km × 1 km grid squares in England (29 'populations' at 13 sites) and Wales (four sites). Research has led to a good understanding of the species' requirements. Its recent reintroduction to a site in Devon appears to have been successful. It is protected under both European and domestic legislation (see *pages 212–214*).

WHERE TO LOOK

The main sites are in the New Forest, Mynydd Preseli and the Itchen Valley. Smaller colonies are found in Purbeck, the Test Valley, Devon and Gower.

OBSERVATION TIPS

Flies low and weakly, and only in sunshine during the hottest part of the day, over or near its favoured streams and flushes.

Males have a distinctive 'Mercury' mark on S2 and spear-shaped black markings on the blue abdominal segments

Females can be confusing and unless caught to examine the pronotum (**which requires a licence**) are best identified by the presence of males!

Northern Damselfly

Coenagrion hastulatum

Eu: Spearhead Bluet

GB Red List (ENDANGERED)

Scottish Biodiversity List

Rare and local, in Scotland

Lake, pond, bog

J F M A M J J A S O N D

Overall length:	31–33 mm
Hindwing:	17–22 mm

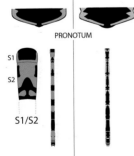

M   F

THORAX

PRONOTUM

S1

S2

S1/S2

ABDOMEN (life-size)

LOOK-ALIKES

Azure Damselfly (*page 86*)
Form *melanogastrum* of Small
Red Damselfly (*page 74*)

A rare species, known in Britain from only a few sedge-fringed lochans in central Scotland.

Adult Identification: Both sexes show the '*Coenagrion* spur' (see *page 37*), which is lacking in Common Blue Damselfly (*page 90*), the only other blue damselfly likely to be found within range (Azure Damselfly (*page 86*) is rare in this area). The eyes and face have bright green undersides, and the pronotum has a blunt point in the middle of the hind margin. **MALE:** On the abdomen, S8–9 entirely blue apart from two small black spots on S9; S2 has variable black 'arrow-head', usually linked to the division between S2 and S3 and detached from a black line along each side; S3 has a black 'spearhead' shape. The lower anal appendages are longer than the upper and form short incurved 'hockey sticks' from above. **FEMALE:** Distinctly green, with top of the abdomen mostly black.

Behaviour: Adults fly more weakly than Common Blue Damselfly, rarely venturing out over open water. They tend to favour sedge beds, usually being found low down close to the water's surface. Eggs are laid, while 'in tandem', into floating and emergent vegetation, including Bog Pondweed and Water Horsetail, the pair sometimes submerging in the process.

Breeding habitat: Small, shallow (<60 cm deep), partly overgrown pools and sheltered margins of lochs with dense beds of emergent sedges and horsetail. The acidic nature of such waters makes them largely fish-free, which possibly allows the larvae to survive in their preferred open stands of emergent vegetation.

Population and conservation: Known in Britain from only about 30 sites in Speyside, Deeside and Perthshire. Populations are typically small, although the species can be locally abundant. Potential threats include natural succession, lowering of the water table due to the effects of afforestation and larval competition from Azure and Common Blue Damselflies. The species has shown an ability to colonise suitable nearby sites (*e.g.* dammed boggy streams) and to survive at sites that dry out temporarily.

WHERE TO LOOK	OBSERVATION TIPS
Can be found in newly created roadside ponds at Abernethy RSPB reserve, Highland, where boggy streams have been dammed, and at the Dinnet National Nature Reserve, Deeside.	Look in and around emergent sedges, Common Reed and Water Horsetail. It has a weaker flight than Common Blue Damselfly and unlike that species rarely flies over open water.

Both sexes have bright green undersides to the eyes and face and the 'Coenagrion spur'

Males have a black 'arrow-head' and detached side-lines on S2

Like Common Blue Damselfly, males have blue S8–9 but with tiny black dots on S9

Irish Damselfly   *Coenagrion lunulatum*

Eu: Crescent Bluet
Ir: Irish Bluet

Irish Red List (VULNERABLE)

W&NE Act (NI) 2011 (Sched. 5)

BAP Priority Species (NI)

Scarce and local, in Ireland

Lake, pond, bog

J F M A M J J A S O N D

Overall length:	30–33 mm
Hindwing:	16–22 mm

THORAX

PRONOTUM

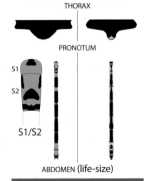

S1

S2

S1/S2

ABDOMEN (life-size)

LOOK-ALIKES

Azure Damselfly (*page 86*)
Variable Damselfly (*page 88*)
Common Blue Damselfly (*p. 90*)

Despite its name, this species was not recorded in Ireland until 1981. It is widespread in northern Europe, occurring as close as The Netherlands, and may await discovery elsewhere in Britain!

Adult Identification: Both sexes show the '*Coenagrion* spur' (*page 37*) and have green eyes and a raised central lobe protruding on the rear edge of the pronotum. **MALE:** The darkest of the blue damselflies occurring in Ireland. The rather dark blue abdomen is mainly black on the top of S3–7 and S8–9 are entirely blue apart from a pair of small black dots on each. The S2 marking is usually an isolated black 'crescent' with a detached black line along each side. The underside is apple-green. **FEMALE:** The top of the abdomen is largely black, like the dark forms of other blue damselflies, but the front end of S8 is blue, interrupted by a black 'spike'. The shoulder stripes are usually greenish.

Behaviour: Adults emerge in daylight 60–90 cm up dead reed stems. They shelter in tall vegetation close to water, the males often flying out to sit on floating leaves. Pairing takes place away from the water, females found by water almost invariably being accompanied by males. Eggs are laid, while 'in tandem', into the stems and leaves of pondweeds.

Breeding habitat: Shallow mesotrophic ponds and lakes with clear water, submerged and floating pondweeds and water-lilies, and open beds of fringing emergent plants. Sites range from slightly alkaline fens to acidic cut-over bogs. Most do not appear to be immediately threatened, but diffuse agricultural pollution is a potential long-term threat to water quality. Bog pools may be threatened by successional change.

Population and conservation: The true range of this species may still be incompletely known, as new sites continue to be found. It has been recorded at 94 sites in 16 counties across the northern half of Ireland, but has already been lost from some, mainly due to eutrophication and lowering of water levels. Large populations exist at the best sites, but most sites support only small populations. Most sites in the Republic of Ireland are not protected by any wildlife conservation designation.

WHERE TO LOOK	**OBSERVATION TIPS**
Montiaghs Moss Nature Reserve, County Antrim, is a well known site in Northern Ireland.	Males fly out to sit on pondweeds and water-lilies more than the other blue damselflies.

Females have extensive blue at the base of S8

Males have green underparts and more black on the abdomen than in other blue damselflies

The 'crescent' and side-lines on S2 are reminiscent of a 'moustache and sideburns'

Dainty Damselfly *Coenagrion scitulum*

Eu: Dainty Bluet

**Rare and local in England
(Red Listed as Regionally
Extinct but recently
discovered breeding)**

Pond, ditch

J F M A M J J A S O N D

Overall length:	30–33 mm
Hindwing:	15–20 mm

M   F

THORAX

PRONOTUM

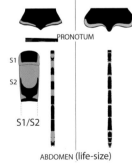

S1

S2

S1/S2

ABDOMEN (life-size)

LOOK-ALIKES

Blue damselflies (*pages 68–81*)

Present in Essex during 1946–52 but lost due to coastal flooding. In 2010 it was discovered breeding in north Kent, since when it appears to have established a toehold.

Adult Identification: A delicate blue damselfly with narrow shoulder stripes and 'Coenagrion spur' (which sometimes has an isolated black dot) on the side of the thorax (see *page 37*). Both sexes show lemon-yellow undersides and pale wing-spots that are longer than in other *Coenagrion* species (nearly twice as long as they are broad). **MALE:** On the abdomen, the S2 marking typically resembles a stalked wine glass; S6–7 appear entirely black; S8–9 are blue (sometimes a little black on S9); and S10 is mostly black. The upper anal appendages are longer than the lower, but both are shorter than in Southern Damselfly (*page 78*).
FEMALE: S3–4 have black 'rocket' shapes, like Common Blue Damselfly (*page 90*), but lacks the broad shoulder stripes and vulvar spine of that species. The shoulder stripes may be yellowish. The rear edge of the pronotum is tri-lobed, like Variable Damselfly (*page 88*), but with a blue spot on each outer lobe.

Behaviour: Males often sit on floating vegetation well offshore, investigating any approaching damselflies. Egg-laying takes place 'in tandem' into surface pondweeds or debris.

Breeding habitat: The early English sites comprised a pond and nearby ditches in coastal grazing marshes with abundant vegetation. The Kent sites are brackish grazing marsh ditches (with Spiked Water-milfoil and Sea Club-rush), and a pond in grassland (with Canadian Waterweed and Rigid Hornwort). In continental Europe, it is found in a variety of similarly well-vegetated water bodies. It is able to tolerate slow-flowing waters and moderate levels of salinity.

Population and conservation: Known from the Hadleigh area of south-east Essex during 1946–52, but not since extensive coastal flooding in early 1953. The species was found on Jersey in 1940–41 and 1950, and rediscovered there in 2009. It was also recorded from Guernsey in 1956. Following a northward spread in continental Europe, it was discovered in 2010 on the Isle of Sheppey, Kent, where small populations have since been found at four sites.

WHERE TO LOOK	**OBSERVATION TIPS**
Coastal ditches and ponds in south-east England with plentiful water-milfoil or hornwort, particularly during early July.	Males often perch on floating vegetation well out over water, and pairs can be seen egg-laying on floating vegetation.

Males have S6–7 completely black and S8–9 blue, with a thick wine-goblet-shape on S2

M

F

F

Both sexes have relatively long, pale wing-spots

M

M

F

Azure Damselfly

Coenagrion puella

Eu & Ir: Azure Bluet

Common and widespread, except in Scotland

Lake, **pond**, river, stream, **canal, ditch**

J F M A M J J A S O N D

Overall length:	33–35 mm
Hindwing:	16–23 mm

THORAX

PRONOTUM

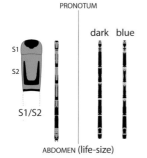

dark   blue

S1
S2
S1/S2

ABDOMEN (life-size)

LOOK-ALIKES

Common Blue Damselfly (*p. 90*)
Variable Damselfly (*page 88*)
Southern Damselfly (*page 78*)

One of the two commonest blue damselflies, being especially at home in small ponds and ditches; most abundant in early summer.

Adult Identification: At close range, both sexes of this species (and all other *Coenagrion* species) can be distinguished from Common Blue Damselfly (*page 90*) by the blue shoulder stripes being narrower than the black lines below them. Also, there is an extra black line (the '*Coenagrion* spur'; see *page 37*) extending from the base of the forewing one-third of the way across the side of the thorax towards the legs. The identification of both sexes can be confirmed by the shape of the hind margin of the pronotum. They lack the pale bar between the eye-spots that is a feature of Variable Damselfly (*page 88*). **MALE:** Characteristic isolated black 'U'-shape on S2 and bow tie shape on the rear part of S9, although there is some variation in these and the other black markings. **FEMALE:** Generally green with extensive black on the abdomen, but a blue form **Fb** occurs in which the blue on the top of the abdomen extends for a third or less the length of each segment on S4–5 (more than one third is blue in the similar form of Variable Damselfly.

Behaviour: Following a fairly concentrated period of emergence in spring, mature individuals are frequently seen mating and laying eggs around the margins of water bodies. The elongated eggs are laid, with the pair 'in tandem', into vegetation at or just below the water's surface.

Breeding habitat: Breeds in a wide range of standing waters, including those that are acidic or eutrophic, but prefers smaller, more sheltered sites. It is regularly found in garden ponds and small ditches, from which the other common blue species, Common Blue Damselfly, is usually absent.

Population and conservation: A common and widespread species of lowland sites, but scarcer in upland areas and in Scotland, where it has recently spread north of the Central Lowlands.

WHERE TO LOOK	**OBSERVATION TIPS**
Around the margins of small ponds and watercourses, or sheltered parts of larger water bodies, particularly in the lowlands.	Rarely ventures far out over large stretches of open water, unlike the other widespread blue damselfly, the Common Blue Damselfly.

Think of the male as a snooker player: he has a cue (the 'spur' on the side of the thorax), wears a bow tie (S10) and carries a beer glass (S2)!

The 'Coenagrion spur' on the side of the thorax and relatively narrow shoulder stripes are characteristic of all the blue damselflies apart from Common Blue Damselfly

Variable Damselfly

Coenagrion pulchellum

Eu & Ir: Variable Bluet

GB Red List (Near Threatened)

Scarce and local

Lake, pond, canal, ditch, bog

J F M A M J J A S O N D

| Overall length: | 33–38 mm |
| Hindwing: | 16–23 mm |

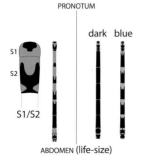

THORAX

PRONOTUM

dark blue

S1

S2

S1/S2

ABDOMEN (life-size)

LOOK-ALIKES

Azure Damselfly (*page 86*)
Irish Damselfly (*page 82*)
Common Blue Damselfly (*p. 90*)

This species has a patchy distribution in Britain, where it is rarer than the Azure Damselfly. However, in Ireland, it is more widespread and often the commoner of the two species. Despite its name, it is not the only species of blue damselfly prone to variations in patterning.

Adult Identification: Both sexes show the '*Coenagrion* spur' (*page 37*) and have a distinctive three-lobed rear edge to the pronotum, the narrow middle lobe of which is more prominent (especially in females) than in Azure Damselfly (*page 86*). They also have a pale bar between the eye-spots, unlike Azure Damselfly. **MALE:** The shoulder stripes are usually broken, sometimes even absent. The abdomen has a thick black 'wine goblet' shape on S2 (occasionally lacking a 'stem'), and S9 is more than half black. **FEMALE:** Occurs in two colour forms: a dark form similar to female Azure Damselfly and a blue form **Fb** which shows a little more blue on the top of the abdomen than Azure Damselfly, with more than one-third of S4–5 being blue.

Behaviour: Most aspects of behaviour are similar to Azure Damselfly, adults often being found in lush vegetation away from the water's edge. In windy weather they may be found sheltering in the lee of bushes. Eggs are laid while 'in tandem' into stems and leaves of aquatic plants, including decaying material.

Breeding habitat: Well-vegetated grazing marsh ditches, fens, ponds, lakes and canals, often in alkaline waters and occurring at the same sites as Hairy Dragonfly and Ruddy Darter. It is sometimes found in cut-over bogs (especially in Ireland) and rarely in slow-flowing waters.

Population and conservation: This species is scarce and very localised in Britain, being inexplicably absent from some apparently suitable sites. Key areas include the grazing marshes of Somerset, Sussex, Kent, the Broads, the Fens, Anglesey and south Galloway. In Ireland, it is more widely distributed, occurring in fens and bogs where it is more common than Azure Damselfly. It is apparently susceptible to frequent ditch dredging, and has been lost from some areas as a result of agricultural intensification.

WHERE TO LOOK	**OBSERVATION TIPS**
In Britain, look along tracks beside well-vegetated ditches in grazing marshes. In Ireland, check small lakes and fens in the midlands.	To check the shape of the pronotum and the pale bar between the eye-spots, try taking and enlarging a digital image.

Beware that although the thoracic stripes are usually 'broken' in males, this is not always the case

M

Think of the male as 'Dracula' – drops of 'blood' drip from 'fangs' (shoulder stripes) into a goblet (S2) and the black mark on S9 vaguely resembles a vampire bat!

M

Blue females have a black marking on S2 reminiscent of a 'Mercury' mark. In dark females, this is 'thistle'-shaped

F

Females are best identified by the pale bar between the eye-spots and the shape of the pronotum

Fb

Fb

Common Blue Damselfly   *Enallagma cyathigerum*

Eu & Ir: Common Bluet

Common and widespread

Lake, pond, **river, canal,** bog

J F M A M J J A S O N D

Overall length:	29–36 mm
Hindwing:	18–20 mm

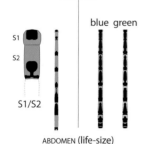

THORAX

PRONOTUM

blue green

S1
S2

S1/S2

ABDOMEN (life-size)

LOOK-ALIKES

Blue damselflies (*pages 78–90*)
White-legged Damselfly (*p. 76*)

The most widely distributed and often most abundant of all the Dragonflies, found in a wide range of habitats with either still or flowing water.

Adult Identification: Both sexes are easily told from other blue damselflies by their broad shoulder stripes and lack of the '*Coenagrion* spur' on the side of the thorax (see *page 37*). **MALE:** Abdomen usually has a black, stalked spot on S2 and completely blue S8–9 (apart from two small black spots on S9). Blue areas fade to grey in low temperatures. **FEMALE:** Occurs as blue **Fb** and dull green **Fg** colour forms, becoming brown with age; the shapes of the black marks on the abdomen include a thistle on S2, various 'rocket' shapes on S3–7 and a triangle on S8. Unlike other blue damselflies, it has an obvious spine in front of the ovipositor, under S8. **IMMATURE:** Pale areas are initially straw-coloured.

Behaviour: A relatively robust damselfly, males and pairs 'in tandem' fly well out over open water in sunny conditions. Eggs are laid, while 'in tandem', into surface vegetation, the male usually releasing the female only when she submerges to continue laying. The males are aggressive when in pursuit of females and patrol the water surface looking for 'damsels in distress'. Struggling to escape the surface meniscus after egg-laying, the females may get a lift back to shore, but the rescuer then expects something in return! In cool or windy conditions, hordes can be disturbed from the relative calm of tall marginal vegetation. Immatures often disperse well away from water.

Breeding habitat: Occurs in a very wide range of waters, both still and flowing, only really shunning small ponds, and can turn up anywhere. Tolerates both nutrient-poor and enriched conditions, but very much dominates the Dragonfly fauna of large eutrophic lakes and reservoirs.

Population and conservation: The most widely distributed and often the most abundant species, with huge populations at many large, shallow standing waters.

WHERE TO LOOK	OBSERVATION TIPS
Easy to find at most sizeable waters, either sheltering around the margins or when characteristically zooming low over the surface well away from the water's edge.	Regularly seen flying out over open water and sitting on floating vegetation. The broad, pale shoulder stripes edged by narrower black stripes are the best field mark to look for.

Fg

M

Remember the phrase "Common Blue is very blue" – the extensive blue tip to the abdomen in the male is reinforced by more extensive blue shoulder stripes and sides to the thorax

M

Mt

M

Fb

Females are readily distinguished by the broad shoulder stripes and lack of a 'Coenagrion spur'

Fg

Fb

Check females for the presence of a spine under S8

Blue-tailed Damselfly *Ischnura elegans*

Eu: Common Bluetail
Ir: Common Bluetip

Common and widespread

Lake, **pond**, river, stream, **canal, ditch,**

J F M A M J J A S O N D

Overall length:	30–34 mm
Hindwing:	14–20 mm

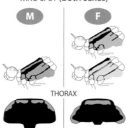

WING-SPOT (both sexes)

M F

THORAX

PRONOTUM

ABDOMEN (life-size)

LOOK-ALIKES

Scarce Blue-tailed Dam'fly (*p. 94*)
Red-eyed Damselfly (*page 96*)

One of our commonest species, easily found in a wide variety of habitats, including brackish and polluted waters.

Adult Identification: Both sexes have two-toned, diamond-shaped wing-spots on the front wings, whitish towards the tip and at least twice as long as wide; abdomen mainly black on top, apart from S8, which is blue or, in some females, brown; and pronotum with distinct lobe in the middle of otherwise straight rear edge. **MALE:** Eyes, thorax, including narrow shoulder stripes, and S8 all blue; thorax greenish in immatures. **FEMALE:** Occurs in five colour forms. Immatures with violet sides to the thorax (*violacea*) mature to have either male-type colouration (*typica*) or an olive-green thorax and brown S8 (*infuscans*). The other immature form (*rufescens*) has orange-pink sides to thorax that become pale brown with a dull brown S8 when mature (*rufescens-obsoleta*): both these forms have only indistinct dark lines below broad, pale shoulder stripes. The wing-spots are less clearly patterned than in the male.

Behaviour: Adults emerge more frequently head-down than other damselflies. Males are territorial. Most adults remain close to water, often congregating in large numbers in marginal vegetation. Individuals may not have attained their mature colouration when they first mate. Mating is prolonged and typically peaks in the afternoon, after which females may egg-lay until the evening. Eggs are laid into submerged plant tissues and debris, without males in attandance. When interacting in flight with other damselflies, this species characteristically jerks rapidly up and down. It is more active in cloudy weather than other damselflies and, despite its weak flight, readily disperses to new locations

Breeding habitat: Found in a very wide range of mainly lowland habitats, although it is less common in acidic waters and avoids fast-flowing waters. It is tolerant of brackish conditions and waters enriched by agricultural run-off and treated sewage effluent.

Population and conservation: Absent from few areas of the British Isles, occurring mainly in the lowlands, and does not require conservation action.

WHERE TO LOOK	**OBSERVATION TIPS**
Can be found almost anywhere in the lowlands among waterside vegetation. Mating pairs are commonly encountered and often allow close examination.	Pairs may spend up to six hours mating, which means that they are more likely to be seen in the 'wheel position' than any other Dragonfly.

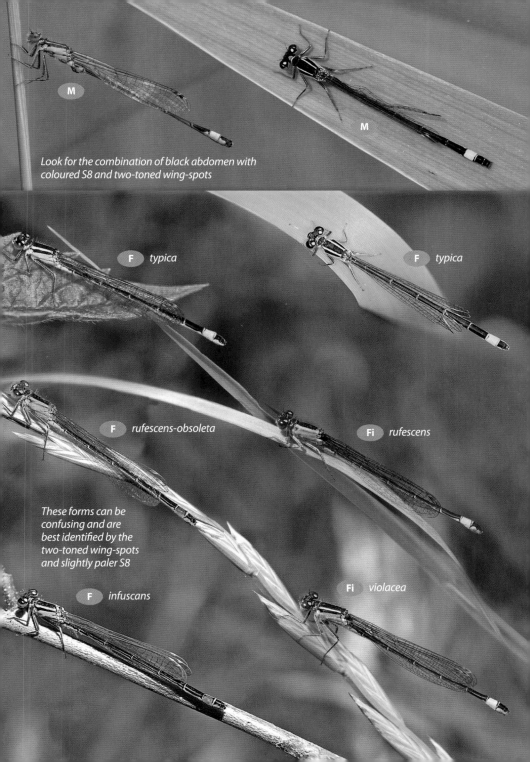

Look for the combination of black abdomen with coloured S8 and two-toned wing-spots

M

M

F *typica*

F *typica*

F *rufescens-obsoleta*

Fi *rufescens*

These forms can be confusing and are best identified by the two-toned wing-spots and slightly paler S8

Fi *violacea*

F *infuscans*

Scarce Blue-tailed Damselfly

Ischnura pumilio

Eu: Small Bluetail
Ir: Small Bluetip

GB Red List (Near Threatened)
Irish Red List (VULNERABLE)
Scarce and local
Pond, stream, flush

J F M A M J J A S O N D

| Overall length: | 26–31 mm |
| Hindwing: | 14–18 mm |

WING-SPOT (both sexes)

M F

THORAX

PRONOTUM

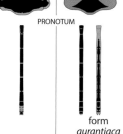

form
aurantiaca
ABDOMEN (life-size)

LOOK-ALIKES

Blue-tailed Damselfly (*page 92*)
Red-eyed Damselfly (*page 96*)

Very localised and rather unobtrusive, favouring sparsely vegetated warm, shallow waters. Easily overlooked, as Blue-tailed Damselfly is usually also present at such sites.

Adult Identification: Both sexes have two-toned, diamond-shaped wing-spots on the front wings, which are shorter than Blue-tailed Damselfly (*page 92*), being < 2× as long as wide, and distinctly larger than those on the hindwings. The rear edge of the pronotum lacks a distinct central lobe. **MALE:** Blue or blue-green eyes, sides of thorax, and narrow shoulder stripes; abdomen mainly black on top apart from blue on S9 and tip of S8, S9 with a pair of tiny, but variable, black marks. **FEMALE:** Greenish eyes and sides of thorax; broad brownish-green shoulder stripes with very narrow black stripe below; abdomen black above; wing-spots browner and less clearly two-toned than male. **IMMATURE:** Coloured parts paler and greener in males; orange in female form *aurantiaca* extends onto wing veins and top of S1–3.

Behaviour: Despite usually flying low and weakly, this species is well able to disperse over long distances by using favourable air-currents. Coastal records may involve immigrants from continental Europe. The female oviposits alone, laying her eggs mainly into emergent grasses and rushes above or just below the waterline.

Breeding habitat: Occurs in small streams, flushes and ponds often associated with heathland, including historical or recent mineral extraction sites. It can tolerate both mineral-enriched acidic and alkaline waters, but these are invariably shallow and usually sparsely vegetated. Ephemeral sites, such as rainwater pools, settlement lagoons and even wheel ruts, may be suitable, but usually only for a limited time.

Population and conservation: The localised distribution of this species reflects habitat availability and shows a south-westerly bias in Britain, although it is more scattered in Ireland. New sites have been found in both semi-natural and artificial habitats, although it has been lost from some due to habitat succession. Sites only remain suitable if vegetation growth is restricted, for example by mechanical operations, livestock trampling or erosion.

WHERE TO LOOK	OBSERVATION TIPS
Best located at small heathland streams and flushes in the New Forest, Purbeck, south-west England and Wales.	Low, weak flight and easily overlooked. Walk slowly through shallow waters, scanning emergent rushes and low vegetation.

Adult females are confusing: look for the suffused brownish-green thorax lacking clear shoulder stripes

F

M

Males have their 'blue-tails' closer to the tip of the abdomen than Blue-tailed Damselfly

M

After egg-laying, females may have a deposit of fine sediment on all or part of their body

F

M

Immature females are a diagnostic orange colour

Fi   *aurantiaca*

Red-eyed Damselfly

Erythromma najas

Eu & Ir: Large Redeye

Locally common

Lake, pond, river, canal

J F M A M J J A S O N D

Overall length:	30–36 mm
Hindwing:	19–24 mm

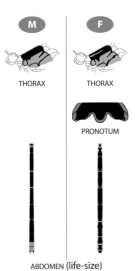

THORAX · THORAX

PRONOTUM

ABDOMEN (life-size)

LOOK-ALIKES

Small Red-eyed Damselfly (*p. 98*)
Blue-tailed Damselfly (*page 92*)

A robust, dark, 'blue-tailed' damselfly, found in the southern half of Britain mainly at large bodies of standing water with floating vegetation, particularly water-lilies.

Adult Identification: Both sexes lack pale spots behind the eyes and have pale brown wing-spots, black legs and two black lines on the sides of thorax, the upper one a *Coenagrion*-like 'spur', sometimes with a detached spot at the end in males. The wings extend more than halfway along S7. **MALE:** Deep burgundy-red eyes; thorax with top completely bronze-black, lacking shoulder stripes, and sides blue; abdomen has blue S1 and S9–10, otherwise dark on top with slight pruinescence. **FEMALE:** Brownish-red eyes; narrow, short or broken yellowish shoulder stripes; otherwise yellowish-green below and dark above with blue divisions to last few abdominal segments. The rear edge of the pronotum is strongly tri-lobed. **IMMATURE:** Like female but the coloured areas are yellowish.

Behaviour: In fine weather, males are found flying low over the water or sitting on floating leaves, where they fight for strategic positions near open water. They quickly move to nearby vegetation when the sun goes in, often landing in trees. Eggs are laid, while 'in tandem', into the stems and leaves of floating and sometimes emergent plants. Egg-laying often takes place underwater, the pair remaining submerged 'in tandem' for up to 30 minutes, after which they float back to the surface.

Breeding habitat: Closely associated with floating leaves, typically water-lilies, but also pondweeds and other floating vegetation. Favoured sites include larger ponds, lakes and flooded mineral workings, canals, main drains and sluggish rivers. It appears to tolerate a wide range of water acidity.

Population and conservation: Locally common, but restricted to southern and central England and the Welsh borders, with outposts in Devon and south Wales. There is evidence of recent consolidation within the main range and of some extension. It may be affected by large-scale dredging operations, but appears to tolerate habitat changes resulting from eutrophication.

WHERE TO LOOK	**OBSERVATION TIPS**
Most easily found at lakes and gravel pits in south-east England and East Anglia.	Use binoculars to check dark damselflies sitting on floating vegetation or flying over open water (as do Common Blue and Blue-tailed Damselflies).

Mt

M

M

Mature males look dark with
a blue tip to the abdomen;
the red eyes are not always
easy to see

F

F

Females may be confused
with infuscans Blue-tailed
Damselfly, but lack a
pale S8 and have
incomplete
shoulder
stripes

Small Red-eyed Damselfly *Erythromma viridulum*

Eu: Small Redeye

Widespread in southern, eastern and central England – recent colonist

Lake, pond, river, canal

J F M A M J J A S O N D

| Overall length: | 26–32 mm |
| Hindwing: | 16–20 mm |

M   F

THORAX   THORAX

PRONOTUM

ABDOMEN (life-size)

LOOK-ALIKES

Red-eyed Damselfly (*page 96*)
Blue-tailed Damselfly (*page 92*)

A much smaller version of Red-eyed Damselfly, but otherwise very similar and easily overlooked. Its recent rapid colonisation of England has been little short of sensational. Numbers peak in July and August, after the peak for Red-eyed Damselfly.

Adult Identification: Like Red-eyed Damselfly (*page 96*), both sexes lack pale spots behind the eyes and have pale brown wing-spots and a *Coenagrion*-like 'spur' on the side of the thorax (see *page 37*), which in males often ends in a dot (rarely in Red-eyed Damselfly). However, the legs are paler. **MALE:** The eyes are brownish-red. The thorax is bronze-black on top, with shoulder stripes that are usually broken, and has blue sides. The abdomen is dark except for blue on the top of S1, S9–10 and the sides of S2–3 and S8. There is a black 'X'-shaped marking on the top of S10. **FEMALE:** The shoulder stripes are complete and, like the sides of the thorax, are yellow, green or blue. The abdomen is similar to that of Red-eyed Damselfly, except that the top of S10 is yellow, green or blue and this colouring extends forward just onto the top and down the sides of S9. The rear edge of the pronotum is rounded.

Behaviour: Mating occurs either on floating plants or at the margins. When perched on floating vegetation, males hold their abdomen slightly upcurved (it is held straight in Red-eyed Damselfly). Eggs are laid, while 'in tandem', into floating plants.

Breeding habitat: Frequents ponds, lakes and ditches with a mosaic of floating mats of pondweed, particularly hornworts or water-milfoils, and algae that cover the surface from mid-summer. It seems to be well able to tolerate brackish conditions.

Population and conservation: First found in Britain at three sites on the Essex coast in July 1999, following range expansions in The Netherlands and Germany. It had become locally abundant in south-east England by 2002, when breeding was finally proven, and has since spread west and north as far as Devon, south Wales and Yorkshire. Further range consolidation and extension seems likely.

WHERE TO LOOK	**OBSERVATION TIPS**
Look on mats of floating algae or pondweed on often unprepossessing ponds. Well-established at Hadleigh Country Park and Lee Valley in Essex, Bluewater in Kent and Priory Country Park in Bedford.	Use binoculars or even a spotting scope to see the diagnostic features, focussing your attention on the tip of the abdomen.

Males have shorter wings and are smaller and more slender than Red-eyed Damselfly **M**

In males, look for a black 'X' on the top of S10 **M**

Look for the diagnostic blue 'wedge' along the side of S8 in males **M**

M

M

The abdomen often curves upwards at the tip

Females have complete shoulder stripes and patches of blue on S9–10

F

F

Hairy Dragonfly

Brachytron pratense

Eu: Hairy Hawker
Ir: Spring Hawker

Scarce and local, but increasing

Lake, pond, canal, ditch

J F M A M J J A S O N D

Overall length:	54–63 mm
Hindwing:	34–37 mm

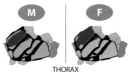

M   F

THORAX

ABDOMEN (life-size)

LOOK-ALIKES

Common Hawker (*page 104*)
Migrant Hawker (*page 106*)
Southern Hawker (*page 110*)
Southern Migrant Hawker (*p. 108*)

The smallest hawker, flying in May and June before the peak emergence of other hawkers. It is found sparingly at unpolluted waters with luxuriant vegetation, mostly in the southern half of Britain and across much of Ireland.

Adult Identification: A small, rather dark hawker with a distinctive hairy thorax. Best identified by the combination of a pair of relatively small, oval-shaped dots on the top of each abdominal segment; yellow costa; long, thin, brown wing-spots; and long anal appendages. The sides of the thorax are extensively green, lacking the pattern of paired pale stripes characteristic of most other hawkers. **MALE:** Blackish, with blue abdominal markings and strong yellowish or greenish shoulder stripes. Lacks the distinctive acutely angled rear margin of the hindwing typical of other male hawkers. **FEMALE:** Similar to the male, but has a hairy abdomen with yellow markings and very restricted shoulder stripes.

Behaviour: Males patrol territories at low level through and alongside emergent vegetation, seeking females and seeing-off intruders. Mating generally takes place in vegetation close to the water body. The female is rather secretive, usually only visiting water to find a mate or lay eggs. These are usually laid into floating dead rushes or other decomposing vegetation, sometimes attended by a male.

Breeding habitat: Found along ditches, canals, ponds and lakes with clean but still or very slow-flowing water and diverse and abundant fringing emergent vegetation.

Population and conservation: This species has been extending its range in Britain in recent years, probably in response to warmer summers and the maturing of open waters such as gravel pits. The management of watercourses to maintain open water with lush fringing and floating vegetation is important for its conservation.

WHERE TO LOOK	**OBSERVATION TIPS**
Look along well-vegetated ditches, such as those running through the grazing marshes in Kent, Sussex, the Somerset Levels and south Wales, and at pools, lakes and coastal wetlands in Ireland.	The only small hawker flying in spring and early summer; on sunny days, look for males patrolling their territories low over the water in and out of emergent vegetation.

The long, brown wing-spots and earlier flight period should avoid confusion with Migrant Hawker, which has a similar abdomen pattern and long anal appendages.

Azure Hawker

Aeshna caerulea

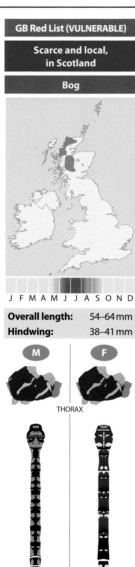

GB Red List (VULNERABLE)

Scarce and local, in Scotland

Bog

J F M A M J J A S O N D

Overall length:	54–64 mm
Hindwing:	38–41 mm

M   F

THORAX

ABDOMEN (life-size)

An early flying hawker, frequenting wooded moorland with bog pools. In Britain, it is only found in Scotland, the Highlands being its stronghold.

Adult Identification: A medium-sized hawker, only overlapping in range with Common Hawker (*page 104*). The head is relatively small, and the eyes touch for only a short distance. The pale stripes on the side of the thorax are wavy and very narrow. The costa is brownish and never bright yellow as in Common Hawker. The brightness of the abdominal markings fades in cool conditions. **MALE:** The abdomen is extensively blue, with paired dots; the lack of yellow markings is diagnostic. The shoulder stripes are very short and the eyes are blue. **FEMALE:** Has blue and brown colour forms that occur with equal frequency; both lack shoulder stripes and have brown eyes. The abdominal pattern has smaller dots than the male; in the brown form the markings are buffish, with small yellowish triangles.

Behaviour: In calm, sunny conditions, males are very active, flying low over moorland and investigating bog pools in search of females. Unlike Common Hawker, they sometimes settle for short periods when searching pools. In cool conditions, both sexes bask often on pale rocks or tree trunks in sheltered locations. Mating occurs away from water and pairs in the 'wheel position' can be found resting on rocks or low vegetation. The female subsequently oviposits alone, laying her eggs into soft substrates or bog-mosses.

Breeding habitat: Breeds in shallow bog pools, often very small, with abundant bog-mosses on moorland up to about 600 m altitude. Adults often shelter and feed in nearby wooded areas, particularly glades in birch woods, and along streams.

Population and conservation: Fairly widespread in the Highlands of Scotland and also occurs in Galloway, although it is rarely found in large numbers. Afforestation and moorland drainage are threats to its conservation, and drought may cause mortality due to the drying out of the shallow pools favoured by this species.

LOOK-ALIKES

Common Hawker (*page 104*)
Hairy Dragonfly (*page 100*)
Migrant Hawker (*page 106*)

WHERE TO LOOK

Search around areas of boggy moorland at Silver Flowe NNR in Galloway, and Glen Affric, Loch Maree and Forsinard RSPB Reserve in Highland.

OBSERVATION TIPS

The best chance of seeing this species is on cool days when both sexes bask in the sun in sheltered locations such as along woodland edges and streams.

M

F

Easily confused with Common Hawker, but smaller with narrower, wavy stripes on the sides of the thorax and brownish rather than bright yellow costa

M

Females lack shoulder stripes

F

In males, the abdomen is extensively blue, lacking any yellow, and the shoulder stripes are very short

F

Common Hawker

Aeshna juncea

Eu & Ir: **Moorland Hawker**

Common and widespread; local in eastern England

Lake, pond, bog

J F M A M J J A S O N D

Overall length:	65–80 mm
Hindwing:	40–47 mm

M **F**

THORAX

ABDOMEN
(life-size)

A large, powerful but wary hawker, most common in late summer in upland areas of northern and western Britain and in Ireland, flying well into autumn.

Adult Identification: Identified by the combination of paired dots on each abdominal segment and the bright yellow costa. The head is large, and the contact between the eyes is broad. The pale stripes on the sides of the thorax are broad and the wings are clear, although some old individuals can show a brownish suffusion. In flight, the abdomen is usually held straight and slightly up-tilted (see *page 56*). **MALE:** Rather dark, with pairs of blue dots and yellow flecks along the abdomen. The yellow shoulder stripes are narrow but quite prominent. The abdomen is noticeably 'waisted'. The eyes are blue. **FEMALE:** Brown, with similar abdominal pattern to the male but the spots are usually yellow, sometimes green, or rarely blue **Fb** (particularly in some Scottish populations); the abdomen is not noticeably 'waisted'. The shoulder stripes are restricted or absent. The eyes are green-brown.

Behaviour: Males are very active, even in overcast conditions, flying low over the water and investigating the edges of pools in search of females. Mating often lasts for an hour or so, and the female subsequently oviposits alone, into aquatic vegetation, detritus or mud, typically above the waterline. Both sexes often feed high up in the open or along woodland rides well away from water, sometimes flying until dusk. They rarely settle in the open.

Breeding habitat: Breeds mainly in acidic standing waters ranging from lakes to boggy pools on moorland and heathland up to about 600 m in altitude.

Population and conservation: Common and widespread in northern and western Britain and in Ireland, but only found locally in eastern England. It may have declined in some areas due to habitat loss.

WHERE TO LOOK	OBSERVATION TIPS
Moorland in northern and western Britain and Ireland. The best places to look in south-eastern England are the New Forest and wet heaths to the west of London.	Look for males over bog pools, or listen for the whirring wings of ovipositing females. Unlike Southern Hawker, does not investigate humans.

LOOK-ALIKES

Migrant Hawker (*page 106*)
Southern Hawker (*page 110*)
Azure Hawker (*page 102*)
Hairy Dragonfly (*page 100*)

Look for the yellow costa, which can even be seen in flight, and the paired spots along the abdomen

Migrant Hawker

Aeshna mixta

Ir: Autumn Hawker

Common in the south; regular migrant

Lake, pond, river, canal, ditch

J F M A M J J A S O N D

Overall length:	56–64 mm
Hindwing:	37–41 mm

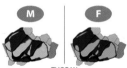

THORAX

ABDOMEN (life-size)

LOOK-ALIKES

Common Hawker (*page 104*)
Southern Hawker (*page 110*)
Hairy Dragonfly (*page 100*)
Southern Migrant Hawker (*p. 108*)

A relatively small hawker that flies well into autumn. It breeds in southern Britain, mainly in standing waters, but also occurs as an autumn migrant.

Adult Identification: Similar to Common Hawker (*page 104*) in having paired dots on each abdominal segment, but the costa is brown, rather than yellow, and there is a narrow yellow triangle on the top of S2. The yellow stripes on the sides of the thorax are broad and the wings are clear (although very slightly tinted in females). In flight, the abdomen is held just above horizontal and appears slightly droop-tipped (see *page 56*). The anal appendages are very long, at least as long as S9 & S10 combined. **MALE:** Appears quite dark, with pairs of blue dots and yellow flecks along the abdomen. Shoulder stripes are faint or absent and the eyes are blue. **FEMALE:** Brown, with similar abdominal pattern to the male, but the spots are smaller and yellow, rarely blue. The shoulder stripes are very restricted or absent. The eyes are green-brown.

Behaviour: Males often fly low over the water in some numbers, hovering frequently and investigating marginal vegetation in search of females. Mating is prolonged, although the female subsequently oviposits alone, laying her eggs into living aquatic vegetation, often well above the waterline. Both sexes frequently perch or hawk high up along hedgerows or woodland rides well away from water and fly until dusk, sometimes in 'swarms'. They regularly bask low down on vegetation.

Breeding habitat: Breeds in well-vegetated lakes, ponds, gravel pits, reservoirs and canals, avoiding acidic waters but tolerating brackish conditions.

Population and conservation: This species has spread northwards and westwards in England, Wales and Ireland in recent decades, and most recently into southern Scotland. The larvae are unable to tolerate low temperatures and this factor appears to determine its distribution. The breeding population is boosted by migrants from continental Europe in the autumn.

WHERE TO LOOK	OBSERVATION TIPS
Common in southern Britain, particularly in the autumn; most easily seen when they are hawking for insects along sheltered woodland rides or tall hedgerows.	In flight, the abdomen is up-tilted and appears to be slightly droop-tipped. Can often be found basking on low, sheltered vegetation, when it may allow a close approach.

The stripes on the side of the thorax are bold but the shoulder stripes are indistinct or absent

Look from above for the combination of paired spots on the abdomen and yellow 'golf tee' on S2

Anal appendages are long, especially in females

Southern Migrant Hawker

Aeshna affinis

(Blue-eyed Hawker or Mediterranean Hawker)

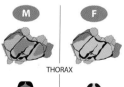

Recent colonist

Pond, ditch

J F M A M J J A S O N D

Overall length:	57–66 mm
Hindwing:	37–42 mm

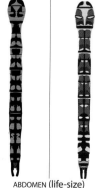

THORAX

ABDOMEN (life-size)

A small hawker that flies somewhat earlier in the year than the similar Migrant Hawker. It was a rare vagrant from mainland Europe until recent influxes and has bred in Essex and Kent.

Adult Identification: A rather small hawker which could be overlooked due to its similarity to Migrant Hawker (*page 106*). However, in both sexes the side of the thorax is essentially pale with only fine black lines and lacks the dark central panel typical of other hawkers. The top of the thorax is brown with short shoulder stripes. The wings are clear, the costa is pale and the wing-spots are ochre-coloured and slightly longer than in Migrant Hawker. **MALE:** The sides of the thorax are green, becoming blue with age. The abdomen is dark with pairs of large bright blue dots along its length and lacks any green or yellow markings when mature; the blue dots on S10 are larger than in Migrant Hawker (see also *page 56* for comparison). The eyes are blue. **FEMALE:** Similar to the male but with yellowish sides to the thorax and smaller paired yellow dots on the abdomen. There is a prominent yellow triangle on S2. The eyes are brown.

Behaviour: Males patrol at a height of about 1·5 m. The flight appears weaker than in Migrant Hawker, with more frequent hovering. It is a somewhat dispersive species, at least in some years. This is the only *Aeshna* hawker that lays eggs 'in-tandem'.

Breeding habitat: Breeds in standing, sometimes slightly brackish, waters with luxuriant emergent vegetation that often dry up during the summer.

Population and conservation: This species has spread northwards substantially across continental Europe in recent years and now occurs just across the English Channel in France. After records from Kent in 1952 and Sussex and Hampshire in 2006, there were multiple records in 2010 near the coasts of Essex, Kent and Norfolk, with egg-laying recorded at Hadleigh Country Park, Essex, and at Cliffe, Kent. Sightings came from eight sites in these counties during 2011–13, with further breeding evidence at Hadleigh. There are also records from Jersey in 1998 and 2004.

WHERE TO LOOK	**OBSERVATION TIPS**
Could occur virtually anywhere, but most likely to turn up at waters near the south or south-east coasts., especially around the Thames Estuary.	Look closely at the thorax of any 'Migrant Hawker' seen before August – if you see one with unmarked sides look closer and try to take a photo or video sequence!

LOOK-ALIKES

Migrant Hawker (*page 106*)
Common Hawker (*page 104*)
Southern Hawker (*page 110*)

From the side, mature males look conspicuously blue, including the sides of the thorax (compare with Migrant Hawker – see page 106)

M

M

M

Mi

Both sexes lack a dark panel on the sides of the thorax

Fi

F

Southern Hawker

Aeshna cyanea

Eu: Blue Hawker

Common in the south, local elsewhere

Lake, pond, canal, ditch

J F M A M J J A S O N D

| Overall length: | 67–76 mm |
| Hindwing: | 43–51 mm |

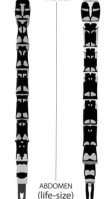

M   F

THORAX

ABDOMEN
(life-size)

LOOK-ALIKES

Migrant Hawker (*page 106*)
Southern Hawker (*page 110*)
Emperor Dragonfly (*page 116*)
Hairy Dragonfly (*page 100*)

110

A large, solitary, inquisitive and colourful species of lowland standing waters, including many garden ponds. Commonest south of a line between the Humber and Ribble estuaries, but recently spreading north into Scotland.

Adult Identification: Both sexes can be identified by coloured bands across S9 and S10 (instead of paired spots as in the other hawkers); very broad, coloured stripes on the top and sides of the thorax; and a narrow yellow triangle on S2. The first two of these features can often be seen in flight (see *page 56*). The wings are clear, with dark wing-veins and wing-spots. There is a black 'T'-shaped marking on the frons. **MALE:** Blackish with apple-green patterning, except for the markings on S8–10 and the side of the abdomen, which are sky-blue (there is a rare form in which all abdominal segments have blue markings). **FEMALE:** Chocolate-brown with green markings. **IMMATURE:** Both sexes are brown with pale yellow markings; the eyes are also brown.

Behaviour: Adults mature away from water, often around tree canopies in woodland glades. Here they hawk insects up to the size of butterflies, sometimes in association with other species. Feeding may continue until dusk on warm evenings. Mature males defend territories aggressively, flying above the water at a height of about 1 m. Females lay their elongated eggs into rotting vegetation, including timber, often above the waterline.

Breeding habitat: Usually found in well-vegetated neutral or alkaline standing waters, sometimes shaded by trees.

Population and conservation: Common in the lowlands of southern Britain, but less so in northern England and scarce in Scotland, where it has spread significantly in recent years. There has been much range consolidation in Wales and in the west and north of England, but intriguingly there has only ever been one record for Ireland (in 1988).

WHERE TO LOOK	**OBSERVATION TIPS**
Almost any non-acidic water, even quite small garden ponds, in the south. Territorial males are inquisitive and may come to investigate you!	Look for 'headlights' on the top of the thorax (shoulder stripes), the coloured bands across S9 and S10, and the chunky thorax. The abdomen is held straight and more or less horizontal in flight.

Males are the most colourful hawkers, with extensive blue and apple-green markings

M

M

F

Fi

Both sexes have 'headlights' and 'tail lights': very broad stripes on the top of the thorax and complete pale bands across S9–10

Brown Hawker

Aeshna grandis

Ir: Amber-winged Hawker

Common and widespread in the lowlands; absent from northern and western Britain

Lake, pond, river, canal, ditch

J F M A M J J A S O N D

Overall length:	70–77 mm
Hindwing:	41–49 mm

ABDOMEN (life-size)

LOOK-ALIKES

Norfolk Hawker (*page 114*)

A large, distinctive hawker with golden-brown wings that breeds in a range of standing waters in the lowlands and is often found away from water.

Adult Identification: Readily identified by its mainly brown appearance, with amber-tinted wings and pair of prominent yellow stripes on the sides of the thorax. Both sexes lack shoulder stripes. **MALE:** Uniformly brown with small blue dots along the side of the abdomen and on S2, and blue on the top of the eyes. **FEMALE:** Like the male but with yellow or pale blue markings along the side of the abdomen and yellowish-brown eyes.

Behaviour: Patrols over water bodies and adjacent emergent vegetation but is often found away from water, sometimes in large numbers. In calm conditions, this species typically flies at canopy level and is often active late into the evening. It has a distinctive flight action that involves long glides interspersed with bursts of rapid but rather shallow wing-beats. Individuals often settle for brief periods to eat prey. Unlike the other hawkers, males rarely clash. Mating is rather prolonged, females subsequently egg-laying alone. The eggs are laid into either living or dead plant material, near or below the water's surface. Emergence takes place at night and adults takes their first flight before dawn.

Breeding habitat: Found in a range of standing or slow-flowing water bodies, including ponds, lakes, gravel pits, canals and ditches. It is able to tolerate moderate levels of pollution.

Population and conservation: This species has probably increased in Britain in relatively recent times due to the creation of canals and gravel pits. Although there has been some consolidation within its core range and slight extension, this may be a result of increased recording effort in recent years. The population is sometimes boosted by the arrival of immigrants from continental Europe in the autumn. It is inexplicably absent from parts of south-west England and Wales, as it is from western France, which makes its widespread presence in Ireland all the more puzzling.

WHERE TO LOOK	**OBSERVATION TIPS**
May be found at virtually any large body of standing or slow-flowing water in the lowlands of Ireland and central, southern and eastern England.	Often feeds high up along sheltered woodland edges or rides, or overgrown hedgerows. It is readily identified by its brown wings and abdomen.

F

F

The wings of both sexes have an unmistakable amber suffusion

M

M

Norfolk Hawker

Aeshna isosceles

Eu: Green-eyed Hawker

| GB Red List (ENDANGERED) |
| W&C Act 1981 (Sched. 5) |
| NERC Act 2006 (S41) |
| UK BAP Priority Species |
| Rare and local |
| Ditch |

J F M A M J J A S O N D

| Overall length: | 62–67 mm |
| Hindwing: | 39–45 mm |

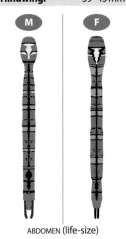

ABDOMEN (life-size)

LOOK-ALIKES

Brown Hawker (*page 112*)

A large, relatively early flying brown hawker. It is found mainly in ditches through grazing marshes and fens in The Broads.

Adult Identification: A large, brownish hawker, lacking any blue, green or yellow markings on the side of the abdomen. The sexes are similar, having brown abdomens with a distinctive yellow triangle on S2, and green eyes. The wings are essentially clear, except for a small orange-yellow area at the base of the hindwings, and amber wing-spots; like other *Aeshna* hawkers, the rear edge of the hindwing is angular in males. Both sexes have a pair of pale yellow stripes on the side of the thorax, but these are less prominent in females. Shoulder stripes are at most faint, but usually absent.

Behaviour: Males patrol low over grazing marsh ditches, often clashing and chasing each other. They spend more time perched than other hawkers. During maturation, which takes up to three weeks, immatures inhabit sheltered woodland edges and paths through tall herb fen. Copulation usually takes place on emergent vegetation. Females oviposit alone, making short flights between egg-laying sites. The eggs are laid, usually below the water's surface, into living Water-soldier plants and rarely into other species or dead plant material. There are two periods of emergence, the first mainly males and the second mainly females. Emergence is typically on Water-soldier leaves.

Breeding habitat: Confined to well-vegetated grazing marsh and fen ditches with clear water and stable water levels, usually with at least 70% cover of Water-soldier. Recent records outside The Broads include reedy ponds, more like sites in mainland Europe where it is less dependent upon Water-soldier.

Population and conservation: Very rarely seen outside a few sites in the Norfolk Broads and north-east Suffolk. Although afforded considerable legislative protection, it is at risk from habitat change resulting from eutrophication of, or saline intrusion into, freshwater dyke systems, large-scale ditch-clearance operations, fluctuations in water levels and a breach in sea defences. Encouragingly, there have been recent sightings in north Kent, west Norfolk and Cambridgeshire, where many exuviae were found in 2013 on introduced Water-soldier.

WHERE TO LOOK	**OBSERVATION TIPS**
Good places to visit are the Norfolk Wildlife Trust Upton Broad & Marshes and Strumpshaw Fen RSPB nature reserves.	Males often perch on ditch-side vegetation for long periods. Approach slowly and avoid sudden movements to get close views.

Although the green eyes may be detected in flight, close views of a perched individual are needed to see the yellow triangle on S2

F

F

Any 'brown' hawker with clear wings seen in East Anglia in early summer is sure to be this species

M

M

Emperor Dragonfly

Anax imperator

Eu & Ir: Blue Emperor

Common and widespread in southern Britain; scarcer in the north, but increasing

Lake, pond, river, canal

J F M A M J J A S O N D

Overall length:	66–84 mm
Hindwing:	45–51 mm

This is Britain's largest dragonfly. It emerges in late spring and is highly territorial, being easily recognised by its bright colouration and seemingly unending flight well above the water's surface.

Adult Identification: Both sexes have an apple-green thorax without any obvious black markings, a yellow costa and brown wing-spots. They often fly with a drooping abdomen (see *page 57*), which has an irregular thick black line down the top of S2–10. **Male:** The abdomen has a green S1 but is otherwise bright blue, although the colour fades in cool conditions. The eyes are blue-green and the wings are clear. **Female:** The abdomen is mainly dull green, sometimes blue **Fb**. The eyes are greenish and the wings become brownish with age. **Immature:** Pale green with brown instead of black on the abdomen.

Behaviour: Territorial males patrol waters for long periods, typically flying far from the shore and higher than other hawkers. They are the dominant hawker and vigorously chase away intruders. They catch and eat prey on the wing, although large prey, up to the size of Four-spotted Chaser and Large White butterfly, is eaten at rest. As with other hawkers, females are often first noticed when egg-laying, as their wings rustle on marginal vegetation. They almost always lay alone, often well offshore, typically inserting their eggs into pondweed just below the water's surface.

Breeding habitat: Inhabits a variety of well-vegetated ponds, lakes, large ditches, canals and slow-flowing rivers. It also breeds in new ponds and can tolerate brackish water.

Population and conservation: This species has expanded its range markedly northwards and westwards in England and Wales in recent years and has been recorded increasingly in Scotland and southern and eastern Ireland.

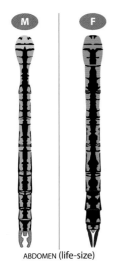

M **F**

ABDOMEN (life-size)

LOOK-ALIKES

Lesser Emperor (*page 118*)

WHERE TO LOOK	OBSERVATION TIPS
Territorial males and egg-laying females can be found readily at many lowland standing waters. Large ponds are often the best places to look.	Any large, powerful, brightly coloured hawker flying before July is likely to be this species. The abdomen is often held down-curved in flight.

M

The green thorax, without black markings, and the black line down the centre of the abdomen eliminate all other frequently encountered hawkers

Fb

F

F

Ir: Yellow-winged Emperor

Rare vagrant, has bred

Lake, pond

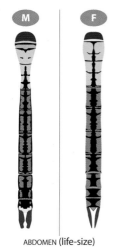

J F M A M J J A S O N D

Overall length:	62–75 mm
Hindwing:	44–51 mm

M F

ABDOMEN (life-size)

LOOK-ALIKES

Emperor Dragonfly (*page 116*)

This large hawker has expanded its range on the near continent and was first recorded in Britain in 1996. Breeding has been proven but it remains to be seen whether it becomes established more widely.

Adult Identification: Slightly smaller than Emperor Dragonfly (*page 116*), with green eyes and violet-brown thorax without any obvious black on slightly green sides (see *page 57*). The wings have a yellowish suffusion on the outer half, a yellow costa and brown wing-spots. The abdomen has an irregular thick black line down the top, brown S1 and a yellow band across the base of S2; rarely, the sides of the abdomen are blue. **MALE:** S2–3 largely bright blue; the black line on the top of the abdomen ends in a point on the waisted S3; the sides of S4–10 are olive-brown or dull blue-green. **FEMALE:** Abdomen usually duller than the male (although sometimes blue **Fb**), with the black central marking extending onto S2; the blue base to the abdomen may be absent. Two tiny protrusions on the back of the head, between the eyes, are diagnostic.

Behaviour: This species has a less powerful flight than Emperor Dragonfly, which is dominant and frequently drives it off. Its behaviour of laying eggs 'in tandem' is unique amongst the hawkers.

Breeding habitat: Typically breeds in ponds and lakes and tolerates brackish water.

Population and conservation: Well over 100 were recorded in Britain in the decade after the first record in Gloucestershire in 1996, followed by a record 90 in 2006. Most have been males and the majority have occurred between mid-June and early September. The records have been surprisingly widespread, including from southern and eastern Ireland, the Isle of Man and Orkney. Breeding was first proved in Cornwall in 1999 and egg-laying has been reported from several scattered sites since then.

WHERE TO LOOK	OBSERVATION TIPS
Could turn up almost anywhere, but recorded most frequently in Cornwall and Kent. The most reliable site seems to be Dungeness, Kent, where the species has been seen regularly since 1998.	The blue base of the male's abdomen shines like a beacon compared with the rest of the body. In flight the abdomen is held straight, without the down-curve typical of Emperor Dragonfly.

M

Fb

Look for the combination of brownish thorax, green eyes and bright blue 'saddle' at the base of the abdomen

A pair of emperors flying or egg-laying 'in tandem' is most likely to be this species

M

F

Golden-ringed Dragonfly *Cordulegaster boltonii*

Eu: Common Goldenring
Ir: Golden-ringed Spiketail

Common and widespread in western Britain; very local in the east

River, stream, flush

J F M A M J J A S O N D

Overall length:	Male:	77 mm
	Female:	83 mm
Hindwing:		41–50 mm

A distinctive, large and impressive dragonfly with black-and-yellow patterning, the female having a long, needle-like ovipositor. It is found along streams and small rivers with acidic running water.

Adult Identification: A large dragonfly. The sexes are similar, with stunning green eyes that meet at a single point. The thorax is striped black and yellow with tapering yellow shoulder stripes. The abdomen is black with alternate broad and narrow yellow 'rings'. The wings are clear, becoming slightly suffused brown with age, and have a yellow costa. **MALE:** The abdomen is swollen at the tip and slightly waisted at S3. The rear margin of the hindwing is acutely angled. **FEMALE:** The abdomen is more parallel-sided than in the male, with a long, pointed ovipositor extending straight out beyond S10. **IMMATURE:** Similar to the mature adult, but the eyes are brown.

Behaviour: Territorial males patrol up and down long beats of suitable breeding waters, their slow flight interspersed with frequent hovering. They are often found with Beautiful Demoiselle and Large Red Damselfly along streams, or with Keeled Skimmer along heathland runnels. They frequently hunt away from the breeding sites, such as among Bracken, heathers and gorse. The eggs are laid while the female characteristically jabs repeatedly down into the bed of shallow water, as if on a pogo stick!

Breeding habitat: Typically breeds in acidic running waters, often in heathland or woodland, where sandy, silty or peaty debris on the bottom provides a suitable hiding place for the larvae. It can be found in waters ranging from tiny boggy runnels to rivers, favouring the quieter stretches where sediment is deposited.

Population and conservation: Widely distributed in western Britain, more locally in the south and north-east. Not recorded in Ireland until two females were found in Co. Kilkenny in 2005, followed by another female in Co. Waterford in 2008.

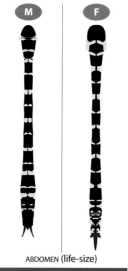

M   F

ABDOMEN (life-size)

LOOK-ALIKES
Female and immature hawkers
(pages 100–110)

WHERE TO LOOK	OBSERVATION TIPS
Easily found during the summer at or near many streams and small rivers in western Britain.	Look for males patrolling low over watercourses. They perch often to eat prey and can then be approached closely. They are inquisitive and may although resembling a giant wasp are of course harmless!

The eyes are emerald green in adults, but brown in immatures

The alternate broad and narrow 'golden rings' across the abdomen are diagnostic.

Fi

F

M

M

Common Club-tail *Gomphus vulgatissimus*

(Club-tailed Dragonfly)
Ir: River Clubtail

GB Red List (Near Threatened)

Scarce and local in southern Britain

River

J F M A M J J A S O N D

Overall length:	45–50 mm
Hindwing:	28–33 mm

A distinctive, medium-sized dragonfly with black and yellow or green patterning, found locally in early summer beside the silty middle and lower sections of a few rivers.

Adult Identification: A medium-sized dragonfly with dull green eyes that do not meet. Both sexes have bold black and yellow or lime-green patterning, which includes two black stripes below very broad shoulder stripes, and discontinuous pale markings down the centre of the abdomen as far as S7. **MALE:** Black and pale green when mature, except for yellow spots on the sides of S8–9; the abdomen has small side protrusions ('auricles') on S2, is 'waisted' at S3 and distinctly clubbed at S8–9; the rear margin of the hindwing is acutely angled. **FEMALE:** the abdomen is thicker and less obviously clubbed than in the male, with broader yellow markings.

Behaviour: Adults emerge synchronously during mornings or early afternoons in mid- to late May. Many emerge horizontally rather than in the more typical near-vertical position. Immatures may disperse up to 10 km. Mature adults are often found in clearings in woodland, scrub and other lush vegetation away from their breeding sites. They often perch in trees, where they can be difficult to observe. Territorial males fly low over the water and return regularly to sit on exposed perches. The spherical eggs are either dropped into the water in flight or dipped onto the surface.

Breeding habitat: Breeds in slow-flowing stretches of a few rivers in southern Britain where silt accumulates, and very occasionally in standing waters nearby.

Population and conservation: Very locally distributed along restricted stretches of a few rivers and some of their tributaries, notably the Dee, Severn, lower Worcestershire Avon, Wye, Tywi, Teifi, Thames (Lechlade to Windsor) and Arun. Threats include water pollution, removal of bankside trees and, during emergence, flooding and wash from boats.

M   F

ABDOMEN (life-size)

LOOK-ALIKES

Black-tailed Skimmer (*page 136*)

WHERE TO LOOK

The muddy larvae can be seen emerging *en masse* at a few places beside certain rivers in late spring. Most easily seen along the Thames at Goring, the Severn at Alveley or even in Shrewsbury, and along the Arun Valley.

OBSERVATION TIPS

Newly emerged tenerals give the best views, but a later trip is needed to see the uniquely black-and-lime-green adult males on territory.

The widely separated eyes are unique in British dragonflies

M

M

Look for the swollen tip to the abdomen and black-and-yellow or black-and- green patterning

F

F

Downy Emerald

Cordulia aenea

J F M A M J J A S O N D

Overall length:	47–55 mm
Hindwing:	31–34 mm

Although very localised, this is the commonest and most widespread of the emerald dragonflies, all of which sport a metallic sheen. It is found at sheltered wooded ponds and flies in late spring and early summer.

Adult Identification: A medium-sized dragonfly that often appears dark unless seen at close range, when its bright green eyes, downy but shiny bronze-green thorax and darker abdomen become apparent. At rest, look for the dark frons with yellowish jaws and golden bases to the otherwise clear wings.
MALE: The abdomen is markedly waisted at S3, with the bulging tip held distinctively above horizontal in flight (see *page 57*); the rear margin of the hindwing is acutely angled. **FEMALE:** The abdomen has almost parallel sides beyond the bulging S2; S9 lacks the conspicuous vulvar scale of Brilliant Emerald (*page 126*). **IMMATURE:** Eyes brown.

Behaviour: Males patrol close to the bank, frequently hovering for a few seconds 0·5–1m above the water in small, sunny bays. They rarely land by water, usually resting in trees some distance away. Adults feed along woodland edges, typically at canopy height. Copulation also takes place away from water and may last for 1–2 hours. The female lays alone by flicking the tip of her abdomen into shallow water to release a gelatinous mass of eggs, often beneath overhanging branches.

Breeding habitat: Usually found at well-vegetated neutral and acidic ponds, lakes and canals fringed with trees. In Scotland and Ireland, it occurs at more open sites with woodland nearby.

Population and conservation: The patchy distribution probably reflects relict populations. It is fairly frequent at wooded, heathland ponds in south-east England, although only small numbers of territorial males are usually present. Threats include the large-scale removal of fringing trees, the dredging of leaf-litter and the introduction of large, bottom-feeding fish.

M	F

ABDOMEN (life-size)

ANAL APPENDAGES

LOOK-ALIKES

Brilliant Emerald (*page 126*)
Northern Emerald (*page 128*)

WHERE TO LOOK	OBSERVATION TIPS
The Moat Pond at Thursley Common, Surrey, is an excellent place to watch territorial males sparring.	Wait for a male to hover briefly in a small 'bay' between trees or in fringing vegetation along the water's edge. Alteratively, scan the margins with binoculars for dark, medium-sized dragonflies that never seem to land.

Look for bright green eyes and bronze-green thorax combined with, in males, a dark bronze 'club-tipped' abdomen, which is elevated in flight (see page 57)

Brilliant Emerald

Somatochlora metallica

GB Red List (VULNERABLE)

Rare and local

Lake, pond, canal

J F M A M J J A S O N D

Overall length:	50–55 mm
Hindwing:	34–37 mm

ABDOMEN (life-size)

S9 S10

ANAL
APPENDAGES

OVIPOSITOR

LOOK-ALIKES

Downy Emerald (*page 124*)
Northern Emerald (*page 128*)

This gem of the emerald dragonflies flies a month or so later than the slightly smaller Downy Emerald. It has an odd distribution, being found locally at neutral or acidic standing waters in Scotland and south-east England.

Adult Identification: A medium-sized dragonfly which, like Downy Emerald (*page 124*), has bright green eyes. However, the whole body is a shining metallic emerald green and the abdomen is longer. The wings have yellow suffusion which is deeper at the bases. The 'face' is like Downy Emerald but has an additional yellow 'U' between the eyes (on the frons) and pairs of small yellow marks on S2–3. **MALE:** The abdomen is waisted at S3 and bulbous towards the tip, but less obviously than in Downy Emerald, widest at S6–7 rather than S7–8, and with the tip held less elevated in flight (see *page 57*). The rear margin of the hindwing is acutely angled. **FEMALE:** The abdomen has parallel sides and a 'spike' (the vulvar scale) is held out conspicuously below S9. The costa is yellow and the rear margin of the hindwing is rounded.

Behaviour: Territorial males patrol the water margins, often under trees, keeping slightly higher and further from the shore than Downy Emerald and not pausing to hover as frequently. They usually venture to nearby trees or bushes to feed, rest or mate. Females oviposit alone, rapidly beating the vulvar scale into shallow water or damp margins with the tip of the abdomen held up.

Breeding habitat: Neutral or acidic ponds and lakes that are at least partially flanked by trees and with woodland close by. Small rivers and canals are used occasionally. The margins of the peaty lochs occupied by this species in Scotland have abundant bog-moss.

Population and conservation: The two disjunct populations in Britain are probably the result of separate post-glacial colonisation events. Threats include large-scale dredging operations and the removal of marginal trees.

WHERE TO LOOK	OBSERVATION TIPS
Small numbers can be found around several woodland ponds in the Thames Valley, such as at Warren Heath in Hampshire. The species is also frequently seen at Coire Loch, Glen Affric in Highland.	As males turn in the sun, look for the diagnostic completely green abdomen, but beware that it can look dark at certain angles.

F

Fi

M

M

M

In sunlight, the eyes, thorax and abdomen of both sexes are, indeed, brilliant emerald!

The male's abdomen is less bulbous than in Downy Emerald, being widest at S6–7

Northern Emerald

Somatochlora arctica

Ir: Moorland Emerald

GB Red List (Near Threatened)

Irish Red List (ENDANGERED)

Rare and local

Bog

J F M A M J J A S O N D

Overall length:	45–51 mm
Hindwing:	30–35 mm

M   F

ABDOMEN (life-size)

ANAL APPENDAGES   OVIPOSITOR

LOOK-ALIKES

Downy Emerald (*page 124*)
Brilliant Emerald (*page 126*)

This peat bog specialist is somewhat intermediate in appearance between the other two emerald dragonflies. Unlike them, however, it typically perches near the ground rather than in trees.

Adult Identification: Smaller than Brilliant Emerald (*page 126*), with a much darker, almost black abdomen. The eyes are bright green and the thorax is metallic bronze-green. Close views reveal a yellow spot on each side of the 'face', next to the eye, and pairs of small yellow marks at the base of the abdomen. **MALE:** The abdomen is narrowly waisted at S3 and bulbous towards the tip (see *page 57*), but less markedly than in Downy Emerald (*page 124*) and widest at S6–7; the anal appendages are distinctive, being large and calliper-shaped from above. The wings have a yellow suffusion, deeper at the bases. The rear margin of the hindwing is acutely angled. **FEMALE:** The abdomen has parallel sides and the short, blunt vulvar scale is held in under S9. The wings have yellow bases and the hindwing has a rounded rear margin.

Behaviour: Males fly low and erratically over bog pools, clashing with other males. Potential breeding sites are often some distance apart and males move frequently between them, sometimes following harassment from Common Hawkers. They forage high among treetops, but mate in lower shrubs. The female lays her eggs alone while in flight, the tip of the abdomen being held up and dipped repeatedly into open water over bog-moss or into wet peat.

Breeding habitat: Breeds in small, shallow bog pools with abundant bog-moss, both at low altitudes and up to 400 m. These are often in or near pine and birch woodland.

Population and conservation: Known mainly from discoveries of larvae or exuviae in its restricted range in north-west Scotland and Killarney in south-west Ireland. Its main threat is habitat loss due to drainage and afforestation.

WHERE TO LOOK	**OBSERVATION TIPS**
Look in boggy areas and clearings beside the A832 at Loch Maree in Highland. In Ireland, try the south shore of the Upper Lake in Killarney National Park.	Fine weather is needed to find adults at the breeding sites.

F

F

In both sexes the abdomen lacks the obvious green hue of Brilliant Emerald

M

M

With close views, the male's narrow 'waist' and distinctive calliper-shaped anal appendages can even be seen in flight (see page 57)

Four-spotted Chaser   *Libellula quadrimaculata*

A rather dull chaser with distinctive dark spots on the wings. It flies during late spring and summer and occurs throughout most of Britain and Ireland.

Common and widespread; irregular migrant

Lake, pond, stream, canal, ditch, bog

J F M A M J J A S O N D

Overall length:	40–48 mm
Hindwing:	32–40 mm

Adult Identification: A medium-sized dragonfly, both sexes having brown eyes, thorax and abdomen. The abdomen is tapered, narrower than Broad-bodied Chaser (*page 132*) and black for the final third or so, with a yellow edge to most segments; it is translucent and irregular polygon shapes can be seen within. Each wing has a yellow base that extends along the front, diagnostic dark spots at the nodes and, like all chasers, extensive blackish patches at the base of the hindwings. The form *praenubila*, which has more black at the nodes and additional dark smudges near the wing-tips, is fairly common. **IMMATURE:** Although the 'four spots' are not apparent on newly emerged adults, they soon darken and the basic body colour becomes bright ochre.

Behaviour: Males are very aggressive and take up territorial perches on emergent or marginal vegetation, from which they sally forth to chase any intruder or intercept a female. Mating occurs in flight and is very brief, lasting only a few seconds. Females that visit water are often harassed and mated by a succession of males. At high densities, territories break down and chaos appears to reign! The female lays her rounded eggs, which are contained in a gelatinous mass, while in flight, by flicking the tip of her abdomen into the water; the male usually hovers protectively nearby.

Breeding habitat: Occupies a wide range of standing waters, from coast to mountains, but the largest populations are associated with acidic heathland pools. It is sometimes found in slow-flowing and even brackish waters and rapidly colonises new sites.

Population and conservation: Although common and widespread, it is less so in north-eastern Britain. It does not appear to be subject to any specific threats.

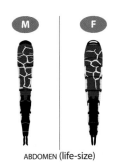

M   F

ABDOMEN (life-size)

LOOK-ALIKES

Broad-bodied Chaser (*page 132*)
Scarce Chaser (*page 134*)

WHERE TO LOOK
Occurs at a wide range of standing waters throughout most of Britain and Ireland and generally easy to find. Wanders widely, often turning up as a migrant on the south coast.

OBSERVATION TIPS
The rather erratic flight, interspersed with bouts of hovering, often attracts attention. It regularly returns to a favoured perch, typically around waist height, where its identity can be easily confirmed.

The dark spots halfway along each wing are diagnostic

M praenubila

M

F praenubila

F praenubila

Broad-bodied Chaser

Libellula depressa

Fairly common in southern Britain; occasional migrant

Lake, **pond,** canal, ditch

J F M A M J J A S O N D

Overall length:	39–48 mm
Hindwing:	33–37 mm

A medium-sized Dragonfly, frequently seen in early summer in southern Britain. It quickly finds and establishes territories at new ponds – sometimes within hours of them being filled!

Adult Identification: Both sexes have brown eyes and thorax, the latter with pale shoulder stripes, and dark patches at the bases of all wings. However, they are characterised by having a short, very broad, flat abdomen with yellow edges to S3–8. **MALE:** Abdomen mostly with powder-blue pruinescence, which can obscure the yellow edges and is typically abraded by the female's legs during mating to produce dark patches; S1 is mostly brown. The shoulder stripes are pale bluish, darkening with age. **FEMALE:** The abdomen is even broader than in the male and yellowish-brown with larger yellow markings on the edges, but may develop limited pruinosity in old individuals; the shoulder stripes are pale yellow. **IMMATURE:** Abdomen is mainly bright ochre, males pruinescing to pale blue.

Behaviour: Males are aggressively territorial, perching on marginal vegetation at between knee- and waist-height. They sally forth to chase intruders or to mate. Their flight is fast, erratic and often difficult to follow. Mating usually occurs in flight and lasts only a few seconds. The female lays her rounded eggs in flight by flicking the tip of the abdomen into the water, usually with the male hovering nearby, although they sometimes visit water when males are absent to avoid being harassed.

Breeding habitat: Prefers small standing water bodies, such as ponds, small lakes and ditches. It is occasionally found at brackish waters but is uncommon at acidic sites. Both well-vegetated waters and those in open, early successional stages are used: it is often the first colonist of new ponds and mineral workings.

Population and conservation: Widespread over much of England and Wales, except the far north, where its range has extended in recent years; a few have also turned up in southern Scotland. It benefits from the creation of new ponds and quickly re-colonises neglected ponds and ditches after dredging.

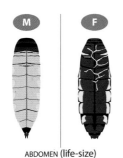

M   F

ABDOMEN (life-size)

LOOK-ALIKES

Four-spotted Chaser (*page 130*)
Scarce Chaser (*page 134*)
Black-tailed Skimmer (*page 136*)

WHERE TO LOOK	**OBSERVATION TIPS**
Found at ponds and small lakes across southern Britain. If you have the time, and live in this part of the country, try digging a pond – then just stand back and wait!	With a careful approach, females and immatures away from water can be viewed closely. Try putting sticks in a pond to tempt males to perch and give you good views.

Females and immatures vaguely resemble a Hornet – and could alarm the unwary!

F

The dark wing-bases and broad abdomen makes this an unmistakable species

M

M

Mi

Scarce Chaser

Libellula fulva

Eu: Blue Chaser

GB Red List (Near Threatened)

Very local, but sometimes abundant

Lake, pond, **river**, canal, ditch

J F M A M J J A S O N D

Overall length:	42–45 mm
Hindwing:	32–38 mm

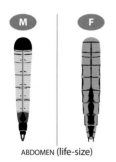

ABDOMEN (life-size)

A very localised early summer species of lush floodplains and washlands, currently expanding its range. Unlike most Dragonflies, immatures are more stunning than adults and are perhaps best appreciated during mass-emergence in May.

Adult Identification: A medium-sized dragonfly. Both sexes have less conspicuous dark wing-bases than the other chasers, but often have dusky patches on the wing-tips. The abdomen is quite broad, especially in females. **MALE:** Eyes blue-grey, whitening with age. The thorax is blackish. The abdomen has a powder-blue pruinescence and a blackish base and tip, resembling Black-tailed Skimmer (*page 136*). **FEMALE:** The eyes are brown. The thorax is olive-brown and the abdomen ochre-brown, darkening and sometimes developing pruinosity with age, with a black mid-line broadening towards the tip into a series of bell-shapes. The wings have yellow suffusion at the front. **IMMATURE:** Like the female, but initially orange-brown.

Behaviour: Males are territorial but can co-exist at higher densities than the other chasers. They regularly chase intruders. Mating occurs while perched and lasts for 15–30 minutes, longer than in other chasers. The eggs, contained in a gelatinous mass, are laid while in flight by flicking the tip of the abdomen into open water; the male usually hovers nearby.

Breeding habitat: Neutral or base-rich, muddy slow-flowing small rivers and dykes in flood-plains and water meadows. It also sometimes breeds in mature ponds, gravel pits and canals. Sites are characterised by dense emergent and marginal vegetation.

Population and conservation: Locally abundant at its favoured haunts on a few river systems in East Anglia and southern England. Some sites have been lost to drainage schemes, but recently its range has expanded in East Anglia and the East Midlands, and new sites have been colonised in Devon, Somerset and River Avon in Worcestershire and Warwickshire. The main threats now include water abstraction and deoxygenation, excessive tree growth and inappropriate cutting of bankside vegetation.

WHERE TO LOOK	**OBSERVATION TIPS**
The drains of the Ouse Washes in Cambridgeshire, the banks of the River Arun in West Sussex and Geldeston Marshes in Norfolk are reliable areas.	Males can often be found perched on emergent plants up to a metre above the water, often where small 'bays' have been created by anglers.

M

Mi

In males, the abdomen is narrower than in Broad-bodied Chaser, but wider and shorter than in Black-tailed Skimmer

Many individuals have diagnostic dark wing-tips

F

Fi

Black-tailed Skimmer
Orthetrum cancellatum

Locally common; increasing

Lake, pond, canal, ditch

J F M A M J J A S O N D

Overall length:	44–50 mm
Hindwing:	35–40 mm

The range of this species has expanded significantly over the last century. At some sites it may be the commonest dragonfly.

Adult Identification: A medium-sized dragonfly with the abdomen tapering evenly from S2–3. The thorax is olive-brown with thin, incomplete black shoulder stripes. The wings are clear with a yellow costa and black wing-spots. **MALE:** The eyes are greenish-blue. The abdomen has powder-blue pruinescence, except for a brown base, black on S7/8–10 and orange arcs on the edges; the pruinescence covers the orange when mature, but may then be abraded during mating. **FEMALE:** The eyes are olive or brown. The abdomen has two black stripes running down the top, forming a ladder pattern; the ground colour becomes dull greyish, sometimes pruinose, with age. **IMMATURE:** The ground colour throughout is bright yellow.

Behaviour: Characteristically perches horizontally on exposed surfaces, such as bare soil and mud, stones, gravel and dead wood. Males protect territories aggressively from exposed perches, frequently flying out low over the water. Mating in flight over water takes barely 30 seconds, otherwise it can last for up to 15 minutes. The eggs are laid while in flight by dipping the tip of the abdomen into water, with the male often hovering nearby.

Breeding habitat: Favours lowland lakes, ponds, sand, gravel, clay and peat workings, large drains and slow-flowing rivers with at least partially exposed margins. It occurs in brackish water and tolerates high fish densities. It is a good colonist of new sites, especially while open ground persists.

Population and conservation: Fairly common in south-east Britain, and abundant at the best sites. Its range has extended northwards in recent decades in response to wetland creation, particularly flooded mineral and peat workings, and a few have reached southern Scotland. In Ireland, it is found mainly at shallow, base-rich loughs in the midlands and the west, but has also spread to some locations in the south and the east.

M   F

ABDOMEN (life-size)

LOOK-ALIKES
Keeled Skimmer (*page 138*)
Scarce Chaser (*page 134*)
Broad-bodied Chaser (*page 132*)
Four-spotted Chaser (*page 130*)

WHERE TO LOOK
Within the main English range, likely to be found basking at any water body with exposed margins. In Ireland, it is found at lakes in The Burren, Co. Clare with similar features.

OBSERVATION TIPS
The flight over water is rapid, just skimming the surface, and jinking characteristically from side to side, with wings raised slightly in a shallow 'V'.

Males lack the dark wing-bases of Scarce and Broad-bodied Chasers and have a longer and narrower abdomen

Females and immatures have a distinctive black 'ladder' pattern on the abdomen and the yellow antenodal cross-veins typical of the skimmers

Keeled Skimmer

Orthetrum coerulescens

Ir: Heathland Skimmer

Scarce and local

Pond, stream, bog, flush

J F M A M J J A S O N D

| Overall length: | 36–45 mm |
| Hindwing: | 28–33 mm |

A rather small dragonfly characteristic of wet heathland, especially in the south and west. The male's pale blue abdomen is more slender than in any similar species.

Adult Identification: Both sexes have a dark brown top to the thorax with buff shoulder stripes (fading with age in males), a thin dark line ('keel') down the top of the abdomen, pale yellow costa and antenodal cross-veins and orange wing-spots. **MALE:** Eyes blue-grey. The evenly tapering abdomen is wholly powder-blue apart from a dark S1. The thorax can become bluish with age, thus resembling the larger Southern Skimmer (*page 172*) (which has yet to be recorded in Britain). **FEMALE:** The abdomen is rich ochre with parallel sides and short dark bars crossing the 'keel' close to segment divisions; it darkens with age and may pruinesce. There is a golden suffusion on the front of the basal half of the wings. **IMMATURE:** Often has a yellowish suffusion to wings.

Behaviour: Males have relatively small territories which they defend from low perches. Mating occurs on the ground and varies in duration from 2–30 minutes or more. The eggs are laid while in flight by dipping the tip of the abdomen into water, with the male often hovering nearby. The flight is fast and erratic, with frequent, brief spells of hovering. At rest, the wings are characteristically held well forward. Immatures sometime disperse from breeding sites and may be found in unsuitable habitat up to 20 km away.

Breeding habitat: Closely associated with acidic wet heathland sites, especially valley mires, frequenting pools, runnels and streams, typically with bog-mosses. It occurs up to about 400 m altitude in south-west England. May also be found at less acidic waters, perhaps with poor breeding success.

Population and conservation: Despite its discontinuous range, it can be locally abundant at sites in southern and south-west England and west Wales. Losses have occurred due to development and peat extraction, but it has shown an ability to colonise suitable new sites quickly.

M | F

ABDOMEN (life-size)

LOOK-ALIKES

Black-tailed Skimmer (*page 136*)
Common Darter (*page 144*)
Black Darter (*page 142*)

WHERE TO LOOK	**OBSERVATION TIPS**
Boggy heathland in the New Forest in Hampshire, Purbeck in Dorset and the fringes of Dartmoor in Devon have good populations.	Look out for males basking low down at the edge of boggy heathland pools and streams. If disturbed from their territorial perches they will often return if you take a few steps back.

M

M

The orange wing-spots and lack of black on the abdomen of males rule out Black-tailed Skimmer

F

F

Females and immatures can be told from Common Darter by their pale shoulder stripes, costa and antenodal cross-veins

Fo

White-faced Darter   *Leucorrhinia dubia*

GB Red List (ENDANGERED)

Scarce and local

Pond, bog

J F M A M J J A S O N D

| Overall length: | 31–37 mm |
| Hindwing: | 23–28 mm |

HEAD ('FACE') (both sexes)

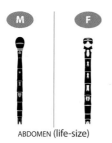

ABDOMEN (life-size)

LOOK-ALIKES

Black Darter (*page 142*)

This highly localised little dragonfly is a spring and early summer specialist of bog pools, now occurring mainly in Scotland, and is almost certainly at risk from climate change.

Adult Identification: Both sexes sport small black wing-bases, a creamy-white frons, dark eyes, black legs, mostly black thorax and abdomen, and dark brown wing-spots, which have white veins radiating out from them. **MALE:** The thorax has red shoulder stripes, two red stripes on the side, and red spots around the wing-bases. The narrow, slightly waisted abdomen is also black, with red markings on the top of S1–7; the red colouration darkens and reduces in extent with age. **FEMALE/IMMATURE:** Patterned like the male, but in yellow, which is more extensive on S1–7 in females; the yellow colouration darkens to ochre or even red with age.

Behaviour: Tenerals disperse to nearby scrub or woodland, which is also used by adults for roosting. Adults spend much of their time perched in vegetation and often bask on pale ground or wood. Mature males are less territorial than other darters, defending relatively small territories. Flies low over the water in a rather erratic manner. Females lay eggs in flight by flicking the tip of the abdomen into waterlogged bog-mosses and between emergent cotton-grass stems.

Breeding habitat: Lowland bog pools with rafts of bog-moss, but also in waterlogged bog-moss depressions devoid of open water. Such waters, often resulting from small-scale peat digging, are too acidic to support predatory fish, which this species cannot tolerate.

Population and conservation: Found mainly in the Scottish Highlands, especially the north-west. It has been lost from half of its English sites in the last 50 years and now occurs at only two sites in the West Midlands and four in Cumbria, including a recent translocation site. Losses have been due to changes in land use, particularly intensive peat extraction, agricultural reclamation and afforestation. However, it remains threatened by habitat succession and climate change.

WHERE TO LOOK	OBSERVATION TIPS
Abernethy RSPB Reserve, Glen Affric and Loch Maree are good Scottish sites. South of the border, it is most easily found at Whixall Moss in Shropshire.	Away from its breeding pools, individuals may be found in and around light woodland and scrub, often basking low down in a sun-trap. Look for its characteristic bouncing flight.

M

M

Ft

The only British species with a pattern of red or yellow on a black abdomen, dark wing-bases, white face and white veins radiating from the wing spots

Fo

F

Black Darter

Sympetrum danae

J F M A M J J A S O N D

Overall length:	29–34 mm
Hindwing:	22–27 mm

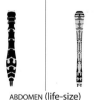

HEAD ('FACE')   M   F

ABDOMEN (life-size)

LOOK-ALIKES

Ruddy Darter (*page 146*)
Common Darter (*page 144*)
White-faced Darter (*page 140*)
Keeled Skimmer (*page 138*)

Britain's smallest dragonfly and a very active late summer species of heathland and moorland bog pools.

Adult Identification: Both sexes have black legs and wing-spots and a very broad base to the hindwing. The frons is pale yellow, darkening with age (see illustration of head). The thorax has yellow sides separated by a bold black panel, within which are three yellow spots. (This feature is also found in female and immature Common Darters (*page 144*) in north-west Britain.) **MALE:** Mainly black on top of the eyes, thorax and abdomen. The abdomen has small yellow marks on the sides (darkening with age), and a distinct 'waist' around S4. The wings are clear. **FEMALE:** Blacker than other darters, with a diagnostic combination of black legs and black triangle on the otherwise brown top of the thorax. The eyes are brown on top. The broad-based, tapering abdomen is yellow-ochre, becoming browner with age, with black inverted 'T'-shapes on S8–9 and broad, continuous black markings down the sides; the vulvar scale is prominent (*page 37*). There are small yellow patches at the wing-bases. **IMMATURE:** As female, but yellower.

Behaviour: The flight action is rather skittish, resembling Ruddy Darter (*page 146*). Males are not strongly territorial, but actively seek out females. Eggs are laid in flight, either 'in tandem' or alone, by dipping the tip of the abdomen into open wet margins, especially where there is bog-moss.

Breeding habitat: Restricted to acidic shallow pools, lake margins and ditches in lowland heathland and moorland blanket bog, usually with abundant bog-mosses and fringing emergent rushes and sedges.

Population and conservation: Very locally distributed in the lowlands, but more widespread in the uplands of northern and western Britain and in Ireland. It is often locally abundant and this perhaps triggers dispersal. Records away from heathland and on the coast result from dispersal or immigration. The main threats come from development, drainage, agriculture, peat extraction and climate change.

WHERE TO LOOK

Look around pools both in lowland mires and higher blanket bogs. It can be found in both habitats on Dartmoor.

OBSERVATION TIPS

Best observed early in the morning or on cool days while they are inactive. To help warm themselves up they often sit on stones or even dry cowpats.

M

Mature black males are readily identified

Mi

Males bask on open ground, but in hot weather sit in the 'obelisk position' to avoid overheating

Fo

Ft

Females and immatures can be confusing: look for the diagnostic dark triangle on the top of the thorax

F

Mi

M

Common Darter

Sympetrum striolatum

(Incl. dark forms previously known as Highland Darter)

Common and widespread; regular migrant

Lake, pond, river, stream, canal, ditch

J F M A M J J A S O N D

Overall length:	33–44 mm
Hindwing:	24–30 mm

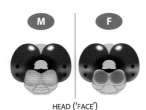

HEAD ('FACE')

ABDOMEN (life-size)

This restless little species is the most widespread dragonfly in the lowlands. It flies late in the year, even into December.

Adult Identification: In both sexes, the thin black line across the top of the frons does not extend down the sides (see illustration of head). The eyes and top of the thorax are brown, the latter sometimes with poorly defined pale shoulder stripes. Like all *Sympetrum* darters, the abdomen has black lines down the middle of S8–9. The legs are brown, usually with a yellowish stripe. The wings are clear with tiny areas of yellow at the base. The wing-spots vary from yellow to reddish-brown. **MALE:** Side of thorax has two large yellow patches divided by a reddish-brown panel. The abdomen orange-red and slightly waisted. **FEMALE:** The abdomen is ochre, becoming very dull, and sometimes reddish with age. **IMMATURE:** Paler than female, with yellowish-brown eyes and thorax. **VARIATION:** Darker in north-west Scotland, and occasionally in other areas, with more extensive black down the side of the eyes and abdomen and on the legs. The darker panel on the side of the thorax has three small, irregular reddish or yellowish spots. Females may also have a black 'T'-shape on top of the thorax.

Behaviour: Basks on sheltered bare ground, allowing it to be active in cool autumn conditions. Territorial males perch on or near the ground and, in autumn, may hover several metres above breeding waters. Mating takes place at rest and lasts 10–15 minutes. Eggs are laid in flight, usually 'in tandem' or with the male hovering nearby, by repeatedly dipping the tip of the abdomen into shallow water. Immigration occurs regularly with warm southerly or easterly airflows and, unusually, may involve pairs 'in tandem'.

Breeding habitat: Occupies a wide range of habitats, including ponds, lakes, ditches, canals, sluggish rivers and even very shallow and brackish waters.

Population and conservation: Occurs across much of the British Isles, but is largely absent from large swathes of upland country from northern England to central Scotland.

WHERE TO LOOK	OBSERVATION TIPS
Found almost anywhere in the lowlands and often away from water. Look for it perched on the ground, on brambles, or even high up on branches.	Darts from perches to see off intruders or catch prey, usually returning to the same spot. Familiarise yourself with this species in case you encounter one of the rarer, migrant darters.

Males have a dark panel on the side of the thorax

M

M

A relatively large darter, males being orange-red with only a slightly waisted abdomen

Both sexes have yellow stripes down the legs

F

F Dark form

Ft

Females and immatures in the north-west (above) have more black and can look very like Black Darter – but lack the black triangle on the top of the thorax

Fo

Fo

A small 'red-and-black' darter that is usually less abundant than Common Darter.

Locally common and increasing; irregular migrant

Lake, pond, canal, ditch

J F M A M J J A S O N D

Overall length:	34–39 mm
Hindwing:	24–29 mm

Adult Identification: Both sexes have black legs and a rather short abdomen. A black line extends a short distance down the sides of the frons (see illustration of head). The front of the thorax is marked with a dark 'T'-shape. There is a tiny area of yellow at the base of each wing; the wings may take on a golden hue with age. The wing-spots are red-brown. **MALE:** The distinctively waisted, blood-red abdomen has a 'clubbed' tip and thick black lines down S8–9. The thorax is unmarked red-brown. The eyes are red-brown on top and dark green below. The frons is red. **FEMALE/IMMATURE:** Thorax is yellow-ochre with thin black lines diagonally across the sides. The abdomen is largely yellow-ochre, with more conspicuous black markings on the sides than on Common Darter (*page 144*). The eyes are brown on top, yellow below. The frons is yellow. Old females may develop red on the abdomen.

Behaviour: Territories are established typically slightly back from the water's edge. Mating lasts a few minutes and occurs usually while perched and away from water. The unusually large eggs are laid in flight, either while 'in tandem' or with the male in attendance. They are deposited into water among dense vegetation or into damp mud.

Breeding habitat: A specialist of well-vegetated ponds, lakes, canals, ditches and, sometimes, sluggish rivers, often with woodland nearby. It also tolerates acidic and brackish waters and, like emerald damselflies, can breed in waters that dry out in late summer.

Population and conservation: Most common in southern and eastern England but generally increasing and expanding northward and westward. It is quite widespread in Ireland. This species disperses or migrates to some degree, sometimes occurring at coastal sites. It is most at risk from excessive dredging, drought, habitat succession and scrub encroachment.

HEAD ('FACE')

ABDOMEN (life-size)

WHERE TO LOOK	OBSERVATION TIPS
Look along ditches or around ponds choked with emergent vegetation. Grazing marshes in East Anglia and the Somerset Levels have large populations.	Wait for them to return to perches on vegetation to get a good view, especially to confirm that the legs are black. The flight is characteristically jerky, with more hovering than Common Darter.

LOOK-ALIKES

Other darters (*pages 142–150*)
Keeled Skimmer (*page 138*)

M

M

Males have a clearly waisted, blood-red abdomen and a reddish head and thorax

The only 'red' darter with completely black legs

Mi

Mi

Mi

F

F

Females and immatures have a 'T'-shaped dark marking on the top of the thorax

Red-veined Darter

Sympetrum fonscolombii

An attractive darter, which has long been known as a migrant, mainly to coastal sites in the south-east, but is now appearing more widely and breeds at some sites.

Adult Identification: Slightly larger than Common Darter (*page 144*). Both sexes have eyes that are blue below; the black line across the top of the frons extends down the sides (see illustration of head). The wings often appear blue-grey and have yellow bases intermediate in extent between Common and Yellow-winged (*page 150*) Darters. The wing-spots are usually pale, strongly outlined in black. There are pale stripes down the legs and the abdomen is virtually parallel-sided. **MALE:** The costa and other veins in the inner half of the wings are conspicuously red. The eyes are red-brown above. The frons and top of thorax are reddish, the latter with a single pale line on the sides. The abdomen is mostly pinkish-scarlet above. **FEMALE:** Extensive yellow veins on the basal half of the wings. The thorax and abdomen are yellow-ochre, the former with dull greenish-yellow sides. The eyes are brown above.

Behaviour: Males are very territorial, spending more time in flight than perched. The flight is strong and erratic, with frequent hovering, often far out over water. Eggs are laid in flight by repeatedly dipping the tip of the abdomen into open water, usually 'in tandem'. Emergence has been recorded both in early summer (June) and later in the summer (August–September), reflecting the two generations typical of southern Europe. This species' inherent tendency to disperse after emergence, especially in late summer, may explain why it does not persist at many breeding sites.

Breeding habitat: Warm, still waters, often bare and shallow, and including brackish sites.

Population and conservation: Mainly an irregular but annual migrant, occurring principally in southern Britain in late summer. The numbers vary from year to year: for example, only 16 migrants were seen in 2004, but a record 900 in 2006, including the first in Scotland for 100 years. Breeding records have increased in recent years and there is evidence that it has persisted at some sites. This suggests that colonisation is underway.

Scarce migrant and rare breeder

Lake, pond, ditch

J F M A M J J A S O N D

Overall length:	33–40 mm
Hindwing:	26–31 mm

M   F

HEAD ('FACE')

ABDOMEN (life-size)

LOOK-ALIKES

Other darters (*pages 142–150*)
Keeled Skimmer (*page 138*)

WHERE TO LOOK	OBSERVATION TIPS
Many sightings have come from well-watched coastal sites, such as bird observatories.	Use binoculars to check the identity of bright red darters flying well out over water: look for the pale stripe on the side of the thorax.

The red wing veins of males are more conspicuous than in other darters

M

M

Fo

Mi

F

The blue undersides to the eyes are characteristic of this species and the very rare Scarlet Darter

F

F

Yellow-winged Darter

Sympetrum flaveolum

Rare migrant; has bred

Lake, pond, bog

J F M A M J J A S O N D

Overall length:	32–37 mm
Hindwing:	23–33 mm

M **F**

HEAD ('FACE')

ABDOMEN (life-size)

LOOK-ALIKES

Other darters (*pages 142–148*)
Keeled Skimmer (*page 138*)

A small, colourful darter with a rather erratic flight, which occurs mainly as an irregular immigrant from the east.

Adult Identification: Both sexes are smaller than Common Darter (*page 144*). The wings have a tangerine suffusion and bright yellow veins over much of the basal half. The wing-spots are usually red or brown (cream in immatures) with thick black margins. Poorly marked specimens may be confused with well-marked Red-veined Darters (*page 148*), but the eyes are yellowish below. A black line extends down the sides of the frons (see illustration of head). The side of the thorax is yellowish and relatively unmarked. The abdomen is virtually parallel-sided with continuous black markings down the lower sides. There are yellowish stripes down the legs. **MALE:** The frons is red, the top of the thorax is reddish-brown and the abdomen is orange-red. **FEMALE:** The top of the thorax and abdomen are yellow-brown. There is an unbroken black line along the side of the abdomen. The yellow in the wings extends beyond the nodes but its intensity fades with age.

Behaviour: The flight is rather weak, fluttering and erratic, with frequent hovering. It perches frequently, often part-way up tall vegetation. Territorial males are fairly aggressive. Mating takes place at rest and lasts about 30 minutes. Eggs are laid, mostly while 'in tandem', among aquatic plants in very shallow water or onto damp mud.

Breeding habitat: Shallow marshy pools, lakes, ditches, bog pools and sometimes seasonally flooded depressions. Prefers slightly acidic waters with extensive tall swampy vegetation.

Population and conservation: An irregular immigrant, mainly to eastern England in late summer. A large invasion in 1995 was one of very few during the last century. Despite breeding subsequently at a few sites, populations did not persist for long, suggesting that Britain's climate is unsuitable for colonisation. About 200 were reported in 2006 in a smaller influx.

WHERE TO LOOK	**OBSERVATION TIPS**
Look in tall grassy areas near the coast in eastern and south-east England after a warm easterly airflow in the summer. Great Yarmouth Cemetery in Norfolk has produced frequent sightings during invasion years.	The fluttering and rather erratic flight action is quite striking. One experienced observer has likened the male's hovering and 'bouncing' flight to a 'fizzing tangerine ball'!

Both sexes typically have
more extensive tangerine
suffusion on the wings
than other darters

Former breeders, vagrants and potential vagrants

This section includes detailed accounts for the two species that formerly bred in Britain but were lost in the 1950s and 1960s (Norfolk Damselfly and Orange-spotted Emerald) and eight species that have occurred as vagrants on only a small number of occasions. It also includes seven species that have not been recorded in Britain but which breed on the near-continent and could conceivably cross the English Channel or North Sea.

Although two species have become extinct in Britain in living memory, no fewer than seven have been recorded for the first time since 1979: Southern Emerald Damselfly, Willow Emerald Damselfly, Common Winter Damselfly, Small Red-eyed Damselfly, Lesser Emperor, Banded Darter and Scarlet Darter. In fact, all but one of these has been recorded for the first time since 1995. The occurrence of these species has coincided with significant northward range expansions of many European dragonflies in recent decades. Some have spread from the Mediterranean region and become established as far north as the English Channel and North Sea coast.

While actions to redress water pollution, especially in rivers, during the closing decades of the last century might explain the range expansions of some species in northern Europe, it is probable that climate change has had a more profound influence. Increases in mean temperatures are likely to have both reduced generation times and allowed some species to breed where previously it had been too cold. Extreme weather events, notably summer droughts and associated hot winds blowing towards Britain from the south or east, also seem to have prompted Dragonfly immigration. One such example is the large influx of darters from the east in 1995: Yellow-winged Darters were a prominent feature and some bred successfully for a year or two before populations died out.

Whereas about 120 species of North American birds have found their way to Britain, only one dragonfly, the Common Green Darner (*page 164*), and two butterflies, the Monarch and the American Painted Lady, are known for certain to have crossed the North Atlantic.

Some migrant birds are known to land on ships at sea, but this is not well documented in Dragonflies. However, a Wandering Glider was noted flying around the wardroom of a British warship for a few days prior to returning to Plymouth from action in Singapore in 1955: it could have alighted on the ship in the Indian Ocean, which is known to be crossed by large number of migrating Wandering Gliders at times. One of the authors has seen an exhausted immature Red-veined Darter on board a ship in the southern part of the Bay of Biscay in late August, in the company of various migrant birds, moths and bats.

Female **Blue Dasher** *Pachydiplax longipennis*

A dead female Blue Dasher was found on a North Sea oil-rig in September 1999, although its provenance is uncertain. Males of this common, sometimes migratory North American 'darter' are powder-blue with yellow wing-bases.

The likely origins of the vagrant species that have occurred in Britain are shown on the map below. A few of these species (shown in red text) have bred successfully in Britain or Ireland.

From the East or South-east
Southern Emerald Damselfly
Yellow-spotted Whiteface
Banded Darter
Vagrant Darter
Yellow-winged Darter

From the West
Common Green Darner

From the South or South-east
Common Winter Damselfly
Southern Migrant Hawker
Lesser Emperor
Vagrant Emperor
River Clubtail
Scarlet Darter
Wandering Glider

The following annotations are used on the maps for vagrant, potential vagrant and former breeding species (*pages 154–186*):

European range
Possible direction of movement
Confirmed record
Former breeding area
? Possible record

Small Spreadwing

Lestes virens

(Small Emerald Damselfly)

Potential vagrant

Lake, pond, ditch

Overall length:	30–39 mm
Hindwing:	19–23 mm

True to its name, this species is slightly more delicate than Emerald Damselfly.

Adult Identification: Similar to the other emerald damselflies (spreadwings), with the typical metallic green upperparts and habit of holding its wings partially spread when at rest. Like Southern Emerald Damselfly (*page 68*), the back of the head is yellow and clearly demarcated from the green top of the head, but the wing-spots are brown, edged at the sides with white. In the subspecies in the northern part of its range closest to Britain, *Lestes virens vestalis*, the narrow shoulder stripes do not extend back to the base of the forewing. **Male:** The eyes are blue, but blue pruinescence is limited to S9–10, around the wing-bases and, when fully mature, under the thorax and on the shoulder stripes. The lower anal appendages are short and straight. **Female:** The ovipositor is short, like Emerald Damselfly (*page 64*) and pale; the sheath is slightly pointed and pale.

Behaviour: Has a tendency to wander, although less so than Southern Emerald Damselfly.

Breeding habitat: Seasonally wet ponds, lakes, marshes and in ditches, in the north-west of its range often at peaty sites with bog-mosses and rushes. Sites are typically surrounded by beds of reeds, sedges or rushes, although adults can sometimes be found in drier grassy or heathy areas.

Population and conservation: NOT RECORDED IN BRITAIN OR IRELAND, but found as close as France and The Netherlands. The northern populations emerge in July and are on the wing through to October.

M   F

S10

S9   S10

ANAL
APPENDAGES

OVIPOSITOR

LOOK-ALIKES

Emerald Damselfly (*page 64*)
Scarce Emerald Damselfly (*p. 66*)
Southern Em'ld Damselfly (*p. 68*)

WHERE TO LOOK
Wet heathlands in south-west England would seem likely sites for potential colonists, given the species' preference for such places on the near continent.

OBSERVATION TIPS
Often found feeding away from water in sheltered heathers, rushes and grasses.

Both sexes have a clear-cut, yellow back of the head and obvious white edges to the wing-spots

M

F

Mature males lack any blue at the base of the abdomen

M

F

F

Common Winter Damselfly

Sympecma fusca

Although as large as the related emerald damselflies, the lack of bright colours in this species means it can easily be overlooked.

Rare vagrant, one record

Lake, pond, ditch

| Overall length: | 34–39 mm |
| Hindwing: | 18–23 mm |

Adult Identification: Both sexes are pale fawn, but darker after hibernation, with dark, slightly metallic markings down centre of the abdomen and top of thorax. Although this metallic colouring recalls emerald damselflies, the wings are held closed when at rest, like other damselflies. The brownish shoulder stripes are as broad as the dark lines below them, which, together with 'rocket'-shaped dark markings on S3-6, give a passing resemblance to a dull female Common Blue Damselfly (*page 90*). The long, brown wing-spots are nearer the wing-tip on the forewings than on the hindwings. **MALE:** Upper anal appendages are pale and pincer-shaped like emerald damselflies (*pages 64–70*). **FEMALE:** Anal appendages are longer than in emerald damselflies and straight, but the ovipositor is shorter.

Behaviour: This and another very similar *Sympecma* species, Siberian Winter Damsel *S. paedisca*, are the only European Dragonflies to overwinter as adults. Adults are found in all months on the continent, emerging from July and some persisting until the following June. From late September, adults hibernate in sheltered areas away from water, occasionally rousing to fly in warm weather. They return to water from March and females lay eggs into floating dead stems of reeds and other similar debris. The larvae develop rapidly during the summer and adults emerge from July onwards.

Breeding habitat: Breeds in standing waters, sometimes brackish, with abundant emergent vegetation where adults gather in spring. In autumn, adults move to sheltered areas and hibernate amongst dry grasses, twigs or woodpiles etc., but sometimes on exposed perches.

THORAX (both sexes)

M   F

S10

S9   S10

ANAL APPENDAGES

OVIPOSITOR

Population and conservation: Recorded once in Britain: in December 2009 a female was discovered inside a house at Tonna, near Neath, south Wales. The location was near a canal and the circumstances did not suggest an accidental introduction. Found throughout southern and central Europe and eastwards, this species has increased in The Netherlands in recent years. It formerly bred on Jersey, the only record since 1945 being in 1997.

WHERE TO LOOK	OBSERVATION TIPS
Might be found in late summer near the south-east coast. In rest of Europe often seen away from water, along sheltered rides in woodland.	An inconspicuous damselfly, both in colour and behaviour, often sitting with its abdomen held tight against a perch.

LOOK-ALIKES

Emerald damselflies (*pp. 64–70*)
Common Blue Damselfly (*p. 90*)

Metallic markings and long anal appendages suggest emerald damselflies, but the wings are held together

The wing-spots are in different positions on the forewings and hindwings

F

F

M

M

Eu: Dark Bluet

GB Red List (REGIONALLY EXTINCT)

EXTINCT IN BRITAIN

Lake, river, stream, canal

Previously known in Britain only from Norfolk, where it was last recorded in 1957. It is an unusually dark damselfly, which at first glance may resemble Blue-tailed Damselfly (*page 92*) or Red-eyed Damselfly (*page 96*) rather than a *Coenagrion* species.

Adult Identification: Both sexes of this slender, delicate species have a distinctive black abdomen with coloured areas at both the base and tip. **MALE:** Lacks shoulder stripes (though rarely there may be four spots). The abdomen is dull blue on S1–3 and S8–9 only; S2 has a black semi-circle at the end and, usually, detached side streaks. It has exceptionally long, incurved, spatulate lower anal appendages, about twice the length of S10. **FEMALE:** The shoulder stripes are green and the pronotum is green at the sides with a prominent pale lobe in the middle of the rear edge. The abdomen is black with blue-green on S1 and the basal halves of S2 and S8.

Behaviour: Males perch on floating leaves, but also fly low but strongly through relatively open beds of emergent vegetation.

Overall length:	31–34mm
Hindwing:	16–20mm

Breeding habitat: Lost from its British sites due to natural succession from open water to dry reedbed and carr woodland. In continental Europe, it favours ponds, ditches and sluggish rivers with beds of low-density reed, sedge and Water Horsetail, tolerating only a limited amount of nutrient enrichment.

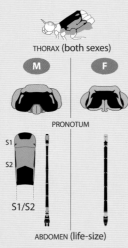

THORAX (both sexes)

M F

PRONOTUM

S1
S2

S1/S2

ABDOMEN (life-size)

Population and conservation: This species acquired its common name from its discovery in 1903 and subsequent presence in a very restricted area of the Norfolk Broads. Records came from Barton, Ranworth, Sutton (Big Bog being the stronghold), Stalham and Hickling Broads. Searches in these areas after the last sightings in 1958 proved fruitless and indicated the continuing unsuitability of the habitat due to the loss of open water or high nutrient levels. In Europe, this species occurs locally from The Netherlands eastwards, but has been lost from many sites. Its main range is now from the Baltic area eastwards. Although believed to have been lost from The Netherlands in the 20th century, it was rediscovered there in 1999.

LOOK-ALIKES

Blue-tailed Damselfly (*page 92*)
Red-eyed Damselfly (*page 96*)

WHERE TO LOOK

Unlikely to be recorded again in Britain, due to its local distibution in north-eastern Europe, and the contraction of its range from the south-west.

OBSERVATION TIPS

The anal appendages of the male are long enough to be visible with the naked eye. The flight period in Britain used to extend from late May to late July.

The female is the only blue damselfly without black on the front half of S2

Males have diagnostic long flipper-like anal appendages and no shoulder stripes

Blue-eye   *Erythromma lindenii*

(Goblet-marked Damselfly)

| Overall length: | 34–39 mm |
| Hindwing: | 18–23 mm |

M   F

THORAX

PRONOTUM

S1

S2

S1/S2

ABDOMEN (life-size)

Like other damselflies, this species has increased in recent years just across the English Channel. Should it manage to make the crossing, however, it could easily be overlooked due to its similarity to Common Blue Damselfly.

Adult Identification: Although it has been re-classified and is now included in the same genus as the red-eyed damselflies, both sexes have shoulder stripes wider than the black stripe below, like Common Blue Damselfly (*page 90*). It has a thin 'Coenagrion spur' on the side of the thorax. The wing-spots are long, pointed and yellow-brown. **MALE:** The eyes are bright blue. The black abdominal markings resemble droplets of viscous liquid dripping down from S1. S7–8 are wholly black, S9–10 are mainly blue, and S10 is split by a black line. The upper anal appendages are long and pincer-like. **FEMALE:** Basically yellow-green, with blue markings on the sides of the central abdominal segments. The dark abdominal markings on S2–7 are narrower than in Common Blue Damselfly and resemble a series of exclamation marks; S9 has a pair of black lines. The anal appendages are yellowish.

Behaviour: Similar to Common Blue Damselfly. Males fly low over open water, perching just above the water on floating vegetation.

Breeding habitat: In continental Europe, this species is found along well-vegetated slow-flowing rivers and canals, and at lakes and gravel pits.

Population and conservation: NOT RECORDED IN BRITAIN OR IRELAND. This species occurs from the Mediterranean northwards, but is scarcer north of the Alps. However, it has increased on the near-continent in recent years and is now locally abundant in northern France and The Netherlands where it flies mainly in July and August.

WHERE TO LOOK	**OBSERVATION TIPS**
In continental Europe, usually found in open water areas or in marginal vegetation, often with Banded Demoiselle (*page 62*), White-legged Damselfly (*page 76*) and other blue damselflies.	Like the related red-eyed damselflies, males are often found on low perches protruding from open water. Look for the mainly blue tip to the abdomen, a feature that is absent in other blue damselflies.

LOOK-ALIKES

Blue damselflies (*pages 78–90*)

Males have S7–8 and most of S6 black, with S2–4 having 'droplet' shapes

M

M

M

M

Both sexes have broad shoulder stripes and a Coenagrion-like spur on the sides of the thorax

Females are dull yellowish at both ends of the body, with blue in the middle

F

F

Vagrant Emperor *Anax ephippiger*

Overall length:	61–70 mm
Hindwing:	43–48 mm

The erratic occurrences of this wanderer from arid lands have often been associated with the arrival of winds from the Sahara.

Adult Identification: A medium-sized hawker, differing from the similar, but larger, Lesser Emperor (*page 118*) in being yellowish-brown with brown eyes. The lower parts of the eyes, thorax and S1–2 are yellow-green. Both sexes have an irregular black marking down the slender abdomen from S3–10, with pale, paired yellowish spots on the top of S8–10. The long, thin wing-spots are brown, the costa is yellow, and the broad hind wings are variably suffused with yellow. Both sexes have pointed tips to the upper anal appendages, which are broader in the female. **MALE:** On the top of the abdomen, S2 is bright blue, producing a more restricted 'saddle' than in Lesser Emperor (see also *page 57* for comparison). **FEMALE:** Darker and duller than the male. The black marking on the top of the abdomen continues onto S2, which is no more than tinged violet.

Behaviour: This species is an opportunistic breeder whose larvae are able to develop rapidly, enabling it to take advantage of transient wet conditions, particularly in hot climates. A compulsive migrant, it may disperse long distances in search of new waters. The species occurs in all months around the Mediterranean, but farther north most records are during July–October. Sightings in Britain and Ireland are often associated with strong southerly airflows originating in North Africa. It may be attracted to lights, including moth traps, when migrating at night.

Breeding habitat: Ponds, lakes and marshes, often temporary and/or brackish.

Population and conservation: Found throughout Africa and eastwards to Pakistan, breeding sporadically in southern Europe and occasionally farther north. More than 20 were recorded in Britain and Ireland prior to an influx of over 20 'hawkers', probably of this species, in early 1998. After singles in October 2010 and January and February 2011, an invasion in Western Europe, assisted by exceptionally hot, southerly winds, resulted in 25 (plus 21 probables) being recorded in Britain and Ireland in spring and at least 12 more in October–November 2011; egg-laying was observed in Cornwall in both April and October. In 2013, 20 or so were seen during July–November, including, in October, an unprecedented eight in Ireland and one egg-laying in Devon.

M | F

ABDOMEN (life-size)

LOOK-ALIKES

Emperor Dragonfly (*page 116*)
Lesser Emperor (*page 118*)

WHERE TO LOOK
Most likely to be found at coastal sites during a hot, southerly airflow.

OBSERVATION TIPS
Check any brown hawker seen in autumn or winter.

Males have an eye-catching blue 'saddle' on the top of S2

M

Both sexes are mainly yellowish-brown and have pairs of yellowish spots on S8–10

F

F

Common Green Darner *Anax junius*

| Rare vagrant |
| Lake, pond |

Overall length:	68–80 mm
Hindwing:	45–56 mm

The commonest North American dragonfly, a migratory species and the first to have made it successfully across the Atlantic to reach European shores.

Adult Identification: Very like Emperor Dragonfly (*page 116*), but both sexes have brownish eyes. Both sexes have a green thorax and a distinctive black and yellow 'bull's-eye' pattern on the 'forehead' (Emperor Dragonfly has a black pentagonal marking). Like Lesser Emperor (*page 118*), the abdomen is brightest at the base, the colour fading quickly towards the tip and with a gradually widening dark pattern. The abdomen becomes purplish during cool conditions. **MALE:** The thorax in front of the forewing bases lacks the blue 'wedges' shown by Emperor Dragonfly. The abdomen has a pale centre to S2 without a black central line. The tips of the long upper anal appendages have tiny spines. **FEMALE:** The sides of the abdomen are brownish or blue-grey, whereas the base is green. **IMMATURE:** The abdomen is red-brown or violet.

Behaviour: This species is on the wing in North America from spring to autumn in the north, and throughout the year in the south. Many are highly migratory, moving north to breed in early spring; the next generation returns south in immature colours in autumn. They migrate during daylight, averaging about 10 km per day. Pairs egg-lay while 'in tandem' (unlike Emperor Dragonfly).

Breeding habitat: Breeds in a variety of standing waters and slow-flowing rivers, including brackish and temporary waters.

Population and conservation: The first records for Europe of this common and widespread North American species involved at least four males and four females in Cornwall and the Isles of Scilly in September 1998. These records coincided with an Atlantic depression associated with Hurricane Earl. In September 2007, a male reached the French coast, near Nantes, Loire-Atlantique.

M **F**

Abdomen
(life-size)

LOOK-ALIKES

Emperor Dragonfly (*page 116*)
Lesser Emperor (*page 118*)

WHERE TO LOOK

Look for vagrants at feeding sites away from water on south-west coasts, after Atlantic storms following an autumn hurricane in eastern North America.

OBSERVATION TIPS

All emperor dragonflies in autumn are worth a close look. If you think you have found this species, try to obtain photographs of the upperparts and particularly the head and base of the abdomen.

The black marking down the centre of the abdomen tapers towards the front and on the male is absent on S2

Both sexes have a round 'bull's-eye' pattern in front of the eyes

NB The photographs shown here are of two of the very few individuals to have made landfall in Britain.

River Clubtail *Gomphus flavipes*

(Yellow-legged Club-tailed Dragonfly)

Rare vagrant, one record

River

Overall length:	50–55 mm
Hindwing:	30–35 mm

The largest of the six European *Gomphus* species, recorded just once in Britain, in the early 19th century.

Adult Identification: Resembles Common Club-tail (*page 122*) but yellower, longer and slimmer, with black-striped yellow legs and a less-clubbed tip to the abdomen. The sexes have similar patterning. The thorax has six thin black stripes on top, the uppermost pair being separated by a central yellow line that widens at the front to produce a yellow 'T' when viewed from above. From the side, these uppermost black stripes join the next ones down, near the front, thereby enclosing two yellow ovals. The abdomen has broader yellow patches on top that extend discontinuously to S8 & S9 (the tops of these segments are black in Common Club-tail). The eyes are blue in mature males, green in females and browner in immatures.

Behaviour: This species has a mid-summer emergence in continental Europe and so the flight period is later than for most clubtails, from mid-June to mid-September, starting a little later in the north. Adults remain close to breeding waters, where they often sit on sandbanks.

Breeding habitat: Breeds in rivers, like most other clubtails, where the larvae live in organic sediment. However, this species favours the sluggish, lower sections of Europe's largest rivers, even as far as the tidal limits. Males patrol territories low over the water's surface, well away from the banks.

Population and conservation: The sole British record is of a male caught at Hastings, Sussex, on 5 August 1818. It is an uncommon species, occurring mainly from eastern Europe into northern Asia. It disappeared from much of western and central Europe but has recolonised some rivers since the 1990s and has been re-discovered in Germany and The Netherlands. The reasons for this recovery are uncertain, but may be linked to improvements in water quality and/or climate change.

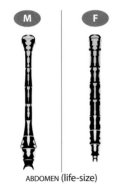

M   F

ABDOMEN (life-size)

LOOK-ALIKES

Common Club-tail (*page 122*)
Western Clubtail (*page 168*)

WHERE TO LOOK
A south-eastern river would seem to be the most probable locality in the unlikely event that this species is recorded again in Britain.

OBSERVATION TIPS
Like other clubtails, most easily found in scrub or grassland away from the riverbank. Territorial males are best looked for by scanning the middle of a river with binoculars.

Both sexes have a yellow 'T'-shape and two yellow ovals on the top of the thorax, and yellow on the legs

The long, slender abdomen has yellow on the tops of S8–9

Western Clubtail

Gomphus pulchellus

(Western Club-tailed Dragonfly)

Potential vagrant

Lake, river, canal

Overall length: 47–50 mm
Hindwing: 27–31 mm

A rather dull yellow-and-black clubtail which is unusual amongst the clubtails in that it is often found at standing waters.

Adult Identification: In both sexes the abdomen has almost parallel sides and lacks the obvious clubbed tip of other clubtails. Like River Clubtail (*page 166*), but unlike Common Club-tail (*page 122*), the legs are yellow and black and the top of the abdomen has a discontinuous yellow stripe down all segments (the tops of S8–S10 are black in Common Club-tail). The top of the thorax has only fine black lines that do not form ovals on top and there are two thin black lines on the sides. The eyes are blue-green.

Behaviour: Males patrol shoreline territories, flying quite fast and low over the water with a rather bouncing action. They rarely hover. Adults frequently settle on bare ground or a low perch a short distance away from the water, where their subdued colours can make them difficult to see.

Breeding habitat: Prefers sluggish rivers and associated standing waters such as ox-bow lakes. In the north of its range, it more often breeds in standing waters such as fishponds, flooded mineral workings and canals.

Population and conservation: NOT RECORDED IN BRITAIN OR IRELAND. This species is endemic to south-west Europe, being found from Portugal to Germany, and has spread north to the English Channel coast in the last century, probably aided by the creation of standing waters. It flies from late May to August in the north of its range.

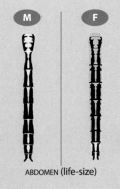

M   F

ABDOMEN (life-size)

LOOK-ALIKES

Common Club-tail (*page 122*)
River Clubtail (*page 166*)

WHERE TO LOOK
A standing water body, such as a sand or gravel pit, small reservoir or fish pond in southern England, seems to be the most likely locality if this species is ever recorded in Britain.

OBSERVATION TIPS
Flies in a rather erratic manner low over the water but often settles on bare ground or low perches near the water's edge. A territorial male is most likely to be seen as it flies along the shoreline.

The side of the thorax has a fine, wavy line down the middle

M

M

There is yellow on the legs and on S8–10 (black in Common Club-tail)

F

The abdomen is almost parallel-sided and the yellow colouration rather faded

Orange-spotted Emerald

Oxygastra curtisii

GB Red List (REGIONALLY EXTINCT)

PROTECTED: EU legislation

EXTINCT IN BRITAIN

River

Overall length: 47–54 mm
Hindwing: 32–36 mm

Although searches since 1963 have failed to rediscover this species in Britain, it may re-colonise or even persist at an undiscovered breeding site.

Adult Identification: Both sexes of this medium-sized dragonfly have a bronze-green abdomen marked with a series of narrow, orange-yellow spots down the centre of all but S8–9. At a distance, it appears dark, but the yellow spot on S10 can be surprisingly obvious. The long, very slender abdomen helps to eliminate Downy Emerald (*page 124*). It also has metallic green eyes and thorax, preventing possible confusion with the clubtails (*pages 122 & 166–168*) or the whitefaces (*pages 140 & 174–176*). **MALE:** Tip of abdomen slightly clubbed; the amber patches at the base of the wings are smaller than in other emerald dragonflies; more angular rear corners to the hindwings than female. **FEMALE:** Front of wings extensively, though variably, suffused amber; base and tip of the abdomen only slightly wider than the rest. **IMMATURE:** Wings may be more extensively amber than females.

Behaviour: Males patrol short, well-spaced beats over slow-flowing, partially shaded stretches of water, flying low and sometimes in a regular, zigzag pattern, raising the abdomen well above horizontal at each change of direction. Usually found close to breeding waters, feeding around trees, often at canopy height.

Breeding habitat: Unlike other emerald dragonflies, this species is associated with muddy sediment mainly in running water. The known English sites were the lower, calm stretches of rivers lined with trees. Slow-flowing rivers and streams, and sometimes canals and ponds, shaded by trees are frequented in continental Europe.

Population and conservation: Known in Britain only from the Moors River and the River Stour, Dorset, during 1820–1963 and the lower River Tamar, Cornwall/Devon, where the sole record is of three (two males) in July 1946. Possible reasons for its extinction from Dorset include pollution (treated sewage), loss of heathland, excessive growth of bankside trees and the cold winter of 1963. Restricted to western European, this species flies from June–August but in England was recorded between mid-June and July.

M   F

ABDOMEN (life-size)

LOOK-ALIKES

Downy Emerald (*page 124*)
Clubtails (*pages 122 & 166–168*)
Whitefaces (*pages 140 & 174–176*)

WHERE TO LOOK

May still persist in quiet stretches of wooded rivers in south-west England or southern Ireland. Could stray across the English Channel.

OBSERVATION TIPS

A very wary species, should you be lucky enough to encounter one! Males patrol territories, particularly during the morning.

Mi

M

When seen from the
side, the abdomen
appears rather sinuous

In flight, look for the male's
very slender abdomen and
broader tip with a pale
'tail light' on S10

F

Southern Skimmer

Orthetrum brunneum

Very similar to Keeled Skimmer and could easily be overlooked. However, it is more akin to Black-tailed Skimmer in many respects, being rather robust, and extensively pale blue in males.

Overall length: 41–49 mm
Hindwing: 33–37 mm

Adult Identification: Both sexes appear generally paler than Keeled Skimmer (*page 138*), have a broader abdomen (the wing-spots are shorter than the width of the abdomen, but longer in Keeled Skimmer) and shorter, reddish-brown wing-spots (2·5–3 mm long, compared to more than 3 mm in Keeled Skimmer). **MALE:** Pale powder-blue pruinescence covers the whole of the thorax and upper surface of the abdomen (Keeled Skimmer has a dark S1 and in old males the thorax may pruinesce, but some evidence of the darker background colour and pale shoulder stripes is usually retained). The 'face' is whitish, becoming pale blue when mature. The eyes are bluish on top. **FEMALE:** The abdomen is brownish-ochre with usually discrete small black spots on each segment, either side of the fine central 'keel' line. The shoulder stripes are usually less conspicuous than in Keeled Skimmer. The 'face' and legs are usually rather uniform and pale. The eyes are blue-green.

Behaviour: Males are territorial, often perching on bare ground, stones or other low perches: behaviour more typical of Black-tailed Skimmer (*page 136*) than Keeled Skimmer.

Breeding habitat: Slow-flowing or sometimes still waters, including small flushes, streams and gravel pits, amid sparse vegetation and bare ground. It is therefore more like Black-tailed Skimmer than Keeled Skimmer in its preference for open habitats.

Population and conservation: NOT RECORDED IN BRITAIN OR IRELAND. Breeds mainly in southern and central Europe, where it flies from June to September (but mainly in July–August in the north). It has spread northwards in recent decades, now reaching the English Channel. It bred on Guernsey in 2000–2001, having perhaps been present there since 1999.

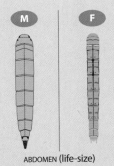

M   F

ABDOMEN (life-size)

WHERE TO LOOK	**OBSERVATION TIPS**
Sites at or near the south coast would seem to provide the most likely chance of finding one.	The venation in the centre of the outer half of the wings is the only certain means of identification. A clear photograph of the outer wing is needed to confirm identification.

LOOK-ALIKES

Other skimmers (*pp. 136–138*)

Males are extensively pale blue with a pale blue 'face'

M

F

M

The double row of at least four cells in the centre of the outer wing is diagnostic (highlighted in red above)

F

Yellow-spotted Whiteface

Leucorrhinia pectoralis

(Large White-faced Darter)

The largest European whiteface, similar to White-faced Darter but with more extensive colouration on the abdomen.

Adult Identification: Larger and more robust than White-faced Darter (*page 140*) and Ruby Whiteface (*page 176*). Compared with these species, both sexes have a broader abdomen, which is swollen around S6, especially in males, and with large coloured areas on the tops of S2–7. The wings have a pale costa and smaller dark patches at the base of the forewings than in White-faced Darter. The wing-spots are very dark brown and have white veins radiating out from them towards the wing-tip. **Male:** The top of the abdomen is yellow, maturing to dull reddish-brown over most of S1–6; there is a diagnostic tapering bright yellow mark on S7, which becomes duller with age, but usually retains some contrast with the other abdominal markings. **Female:** On the abdomen, the yellow on S2 tends to form a band, not an obvious spot as in Ruby Whiteface and the orange-yellow patches on S3–7 are broad. There is a yellow suffusion at the wing bases, which is more extensive in immatures.

Behaviour: Like White-faced Darter, territorial males occupy perches on marginal vegetation. This species is more tolerant of warm conditions than other whitefaces, which perhaps explains its more southerly distribution in Europe.

Breeding habitat: Inhabits boggy pools and lakes, fens, ditches and even slow-flowing rivers and canals. These typically have lush and diverse submerged and emergent plants, such as reedmace and sedges, and are usually mildly acidic. It is able to tolerate a degree of nutrient enrichment at breeding sites.

Population and conservation: The first British record was of one obtained near Sheerness, Kent, in June 1859, but there were no further records until 2012 when males were recorded in Suffolk at Landguard in late May and Dunwich Heath in mid-June. The 2012 records were associated with a major movement across north-west Europe in late spring. It is normally a rather sedentary species that is scarce and local in Europe, its main range being from Germany eastwards. It flies mainly between late May and early July.

Rare vagrant or accidental introduction

Lake, pond, canal, **bog**

| Overall length: | 32–39 mm |
| Hindwing: | 30–33 mm |

M | F

ABDOMEN (life-size)

LOOK-ALIKES

White-faced Darter (*page 140*)
Ruby Whiteface (*page 176*)

WHERE TO LOOK	OBSERVATION TIPS
Could possibly turn up at a bog pool or fen, but is most likely to be found near the east coast. Males often sit on prominent waterside perches.	If you are fortunate to find a vagrant whiteface, concentrate on the abdominal patterning and try to obtain photographic evidence.

A bright spot on S7 usually stands out in males, which have a more 'waisted' appearance than other whitefaces

Females and immature have extensive yellow patches on S2–7

Ruby Whiteface

Leucorrhinia rubicunda

(Northern White-faced Darter)

Larger than the otherwise similar White-faced Darter, with more extensive colouration on a broader abdomen, being more similar to Yellow-spotted Whiteface.

Potential vagrant

Lake, bog

Overall length:	31–38 mm
Hindwing:	27–31 mm

Adult Identification: Slightly larger than the similar White-faced Darter (*page 140*), the abdomen of both sexes having more extensive colour on S2–7. The wings have a yellow costa and smaller dark patches at the bases than in White-faced Darter, especially on the forewings. The wing-spots are reddish-brown and have white veins radiating out from them towards the wing-tip. **MALE:** The red markings on S2–6 fade with age, but the one on S7 remains bright. **FEMALE:** Very similar to Yellow-spotted Whiteface (*page 174*) but the yellow spot on the top of S2 is surrounded by more black than in that species. The yellow markings on the abdomen darken with age. There is a yellow suffusion on the front of the wings that is more extensive on immatures.

Behaviour: Much like White-faced Darter, it perches quite close to the ground and basks in bare places, such as on tree stumps.

Breeding habitat: Similar to White-faced Darter: acidic bogs and lakes with bog-mosses in moorland and woodland. Unlike that species, however, it can occur in waters with at least moderately high densities of fish.

Population and conservation: NOT RECORDED FOR CERTAIN IN BRITAIN OR IRELAND, although up to three out-of-range 'white-faced darters' at Walberswick NNR, Suffolk, in May–June 1992 could conceivably have been Ruby Whitefaces. This is principally a boreal species, occurring eastwards and north-eastwards from Belgium. It can be locally abundant and is said to be common in northern and eastern parts of The Netherlands. However, it has been lost from France and other areas to the south of its current range. The flight period lasts from late April to early August, but is mainly between mid-May and late June in the south-west of its range.

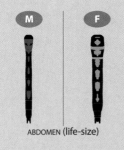

M   F

ABDOMEN (life-size)

LOOK-ALIKES

White-faced Darter (*page 140*
Yellow-spotted Whiteface (*p. 174*)

WHERE TO LOOK	OBSERVATION TIPS
Could possibly turn up at a wetland in eastern Britain, most likely near the coast.	Check very carefully any whiteface away from the few remaining English sites for White-faced Darter. If you are fortunate to find one, focus on the abdominal patterning and try to obtain photographic evidence.

The abdomen is somewhat broader than in White-faced Darter with more extensive coloured markings, but both features are less marked than in Yellow-spotted Whiteface

M

F

Banded Darter

Sympetrum pedemontanum

Pond, ditch

Overall length:	28–35 mm
Hindwing:	21–28 mm

This tiny, but eye-catching, mainly eastern European dragonfly is one of the more unlikely species to have found its way to Britain. It is readily identified by its conspicuous dark wing bands.

Adult Identification: This distinctive species is similar in size to Black Darter (*page 142*) and, like that species, has broad hindwings and black legs. Both sexes have a brown band near the tip of each wing, touching the wing-spots, which are pale in tenerals, becoming yellowish in females and pinkish-red in males. In both sexes, the body colour is much like that of Ruddy Darter (*page 146*). The male's abdomen is slightly swollen-tipped.

Behaviour: The weak, low, fluttering flight is not unlike that of Black Darter. Small numbers wander long distances, although this is a surprisingly unobtrusive species and may be overlooked. It may perch close to the ground or on the tips of sedge and rush stems.

Breeding habitat: Slow-flowing or still waters, including pools, swamps, marshy meadows and drainage channels with fairly dense emergent vegetation and little open water in the west of its range, often in hilly terrain. However, its precise requirements are poorly understood, and in other areas it occupies quarry pools in early successional stages.

Population and conservation: A male in Gwent, Wales, on 16–17 August 1995 coincided with a period of exceptional darter immigration into Britain, apparently originating from the east. This species is very local in central and eastern Europe, although it appears to be spreading westwards. It is sometimes abundant, especially in the east. It flies from July to mid-October in mainland Europe, with numbers peaking in August.

HEAD ('FACE') (both sexes)

M　　　F

ABDOMEN (life-size)

WHERE TO LOOK	OBSERVATION TIPS
Perhaps more than other vagrants, this species might find its way to marshy areas and seepages amid rising ground.	Identification should not prove difficult in this species. It is a weak flyer that tends to keep low down in marshy vegetation and is therefore not easily seen unless disturbed.

LOOK-ALIKES

Other darters
(*pages 140–150 & 180–184*)

Both sexes have unmistakable brown wing bands

M

M

F

Vagrant Darter

Sympetrum vulgatum

Eu: Moustached Darter

Rare vagrant

Lake, pond, river, stream, canal, ditch

Overall length:	35–40 mm
Hindwing:	24–29 mm

HEAD ('FACE') (both sexes)

M   F

ABDOMEN (life-size)

LOOK-ALIKES

Other darters
(*pages 140–150 & 182–184*)

A rare migrant, presumably from eastern Europe, but easily overlooked due to its similarity to Common Darter.

Adult Identification: Very similar to Common Darter (*page 144*), but, as in all the other darters in this book except Common and Southern (*page 182*) Darters, both sexes have a 'moustache': the black line between the eyes extends down the sides of the frons (see *pages 54–55*). The sides of the thorax are poorly marked. **MALE:** The adomen is slightly more waisted and less parallel-sided than in Common Darter. It also tends to be redder and less orange than in that species, and the tiny black spots on S3–7 are surrounded by yellow rings. The thorax is relatively plain in mature individuals, tinged reddish and lacks the two yellow patches on the sides shown by Common Darter. The veins at the front of the forewing are orange. **FEMALE:** Has a distinctive, prominent vulvar scale which projects at a right-angle from under the front of S9. The veins at the front of the forewing are yellowish.

Behaviour: Similar to Common Darter in its habits, favouring low and often pale perches from which to dart up to investigate potential prey or mates. It is an occasional migrant and less prone to large scale dispersion than other darters.

Breeding habitat: A wide range of still and slow-flowing waters, though typically with more luxuriant vegetation than those favoured by Common Darter.

Population and conservation: About ten were recorded up to 1946, mostly in south-east England, with singles in Devon and Yorkshire. However, between 15 June and 1 October 1995, at least 15 were seen in Norfolk (at three sites) and others in Suffolk (two sites), Kent and the London area. This was during an unprecedented immigration of darters of eastern origin (Yellow-winged Darters also occurring in numbers). The only other records are from the Isle of Wight in 1996, the West Midlands in 1997, Devon in 2007 and in a moth trap in Kent in 2013. The species is common in central and eastern Europe, being more abundant than Common Darter in the north of its range. In recent years it has increased in The Netherlands, where hundreds may be seen at times. The flight period in continental Europe is from late June to early November, but it is most abundant from July to September.

WHERE TO LOOK	OBSERVATION TIPS
The south-east coast during a darter influx is clearly the place to look.	Good views and ideally photographic evidence are essential for identification.

Look carefully for a suite of features, including a 'moustache' on males and prominent vulvar scale under S9 on females

Southern Darter *Sympetrum meridionale*

Potential vagrant

Lake, pond

Overall length:	35–40 mm
Hindwing:	25–30 mm

Very similar to Common Darter and easily overlooked. However, careful, systematic checking might turn one up.

Adult Identification: Very similar to Common Darter (*page 144*), but has two black spots on the virtually unmarked sides of the thorax; the uppermost of these spots is very small. The legs are distinctly yellowish-brown, with only a thin black stripe down their length. The wings have black venation, sometimes a little yellow at the bases, and wing-spots slightly larger than in Common Darter. At close range, very little black is visible at the top of the frons, and there is no extension down the sides (see *pages 54–55*). **MALE:** The 'face' is pink. The sides of the thorax and legs may be tinged with red. The abdomen is orange-red and usually lacks any black on S8–9. **FEMALE:** The abdomen is yellowish-brown. The vulvar scale is small and rounded when seen from the side, making it barely visible. The top of the thorax has a suggestion of pale shoulder stripes straddling a dark triangle, somewhat recalling that of Black Darter (*page 142*). The sides of the thorax are pale yellow.

Behaviour: A strong flier that is often found away from water, especially in the south of its range during the hot summer months. It perches on the ground or amongst tall vegetation. This species seems to be particularly prone to infestations of red mites on the wings.

Breeding habitat: Breeds in shallow pools and sheltered parts of lakes with abundant emergent vegetation. Breeding sites often dry out in summer or at least are affected by falling water levels.

Population and conservation: THERE ARE NO BRITISH OR IRISH RECORDS. The validity of four old records is doubtful due to the proven misidentification of one specimen and the likelihood that the others were not of British origin. However, a female was confirmed on Jersey on 5 August 1948. The main range extends around the Mediterranean and eastwards. It also occurs regularly as far north as Brittany and irregularly in hot summers elsewhere on the Channel coast and the southern North and Baltic Seas. The flight period is from July to mid-September in northern France, but from May to early November in the south.

HEAD ('FACE') (both sexes)

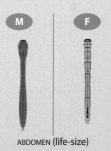

M   F

ABDOMEN (life-size)

LOOK-ALIKES

Other darters
(*pages 140–150, 180 & 184*)

WHERE TO LOOK	**OBSERVATION TIPS**
Most likely to appear in a southern location during a warm, southerly airflow in an exceptionally hot summer.	Look closely at any plain-looking darter, especially one with a heavy infestation of mites on the wings.

Look for two small isolated spots on the side of the thorax and, although not diagnostic, red mites on the wing veins

Fo

Mt

Both sexes have almost unmarked sides to the thorax and largely pale legs

M

F

Males usually lack the small black markings on S8–9 that are found on other darters

M

Scarlet Darter

Crocothemis erythraea

Eu: Broad Scarlet

Rare vagrant
Lake, pond, marsh

Overall length:	36–45 mm
Hindwing:	23–33 mm

Males of this sun-loving species have the most vividly red abdomen of all European dragonflies.

Adult Identification: More robust than other darters, with a flat abdomen that broadens in the middle. There is a large orange or yellow patch at the base of the hindwings and a smaller one at the base of the forewings. The pale wing-spots are outlined strongly in black and are longer than those of other darters. The undersides of the eyes are blue. The sides of the thorax are plain and the legs lack any dark stripes. **MALE:** 'Face' and abdomen are bright scarlet, the latter sometimes with a dark central line on S8–10. The eyes and thorax are reddish-brown. The veins at the front of the wings are red. The legs are orange-red. **FEMALE/IMMATURE:** Dull yellow-brown, becoming olive with age. The shoulder stripes are pale. There is a diagnostic conspicuous pale stripe between the wing bases. The abdomen has a narrow dark line down the centre of S3–10 and may have more-or-less distinct brown stripes down the sides. The vulvar scale protrudes conspicuously, like that of Vagrant Darter (*page 180*). The legs are pale brown.

Behaviour: A strong flier and noted migrant, sometimes moving long distances. It may be attracted to lights at night. Males are territorial at their breeding sites, perching low like other darters with their wings held well forward. There are one or two generations per year in the north of its range and up to four around the Mediterranean.

Breeding habitat: Found in a wide range of shallow, mainly still and even temporary waters, tolerating brackish conditions and moderate levels of nutrient enrichment.

Population and conservation: Seven males have been recorded in southern England since the first in 1995: in Cornwall (two), Cumbria, Devon, Hampshire and the Isle of Wight (two). Records have been in mid-June (three), August (three) and early September. This species occurs commonly in southern Europe, from where it has spread north to reach the English Channel in recent decades. It has bred on Jersey. The flight period is from April to November in southern Europe, but mid-June to September in northern France.

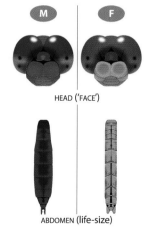

HEAD ('FACE')

ABDOMEN (life-size)

LOOK-ALIKES

Other darters
(*pages 142–150 & 180–182*)

Female skimmers
(*pages 136–138 & 172*)

Oriental Scarlet (*pages 188–189*)

WHERE TO LOOK

Most likely to appear at a sheltered, shallow pond or lake in southern England near the coast.

OBSERVATION TIPS

Could be confused with exotic introduced species (*page 188*), so try to obtain photographic evidence.

Both sexes have a distinctive broad abdomen and lack dark markings on the side of the thorax and legs

Look for the diagnostic conspicuous pale stripe between the wings on females

Wandering Glider

Pantala flavescens

(Globe Skimmer)

Rare vagrant or accidental introduction

Pond

| Overall length: | 45–55 mm |
| Hindwing: | 38–42 mm |

A common tropical migrant that is considered to be the most widespread dragonfly on Earth, having been recorded from all continents except Antarctica. However, it has only occasionally reached Europe.

Adult Identification: Like a large chaser with long wings and exceptionally broad, triangular hindwings that reach almost half-way down the tapering, rather cylindrical, abdomen. There is a yellow patch on the base of the hindwings and males have small brown patches at the wing-tips; the red-brown wing-spots are longer on the forewing than on the hindwing. The abdomen is ochre, with age becoming reddish in males and olive in females, and with an irregular dark central marking that becomes black towards the tip. In flight, the abdomen is typically held down at an angle. The tawny thorax has unmarked yellowish sides. The eyes are brown on top and bluish below.

Behaviour: Usually encountered in flight, which is characteristically high and involves much gliding. It rarely settles but, when it does, it usually hangs vertically. The species frequently feeds in groups, often in swarms when migrating. Migrants at sea may be attracted to ships' lights at night. This remarkable pantropical species probably undertakes the longest migration of any insect on Earth (see box).

Breeding habitat: Breeds in small, often seasonal, pools and sometimes slow-flowing waters, usually with little vegetation.

Population and conservation: The first British record was of one captured in Norfolk, in 1823. Subsequent records from Lancashire in July 1951 and Kent in 1989 may relate to accidental introductions via shipments of imported products (for this reason they are shown as question marks on the distribution map). In 1955, one was found inside a British warship returning from Singapore a few days before reaching Plymouth, Devon. The few other European records all come from south-east Europe, where most sightings have been in late summer or early autumn.

Amazing migrations

The Wandering Glider undertakes truly phenomenal migrations involving vast numbers of individuals that follow the seasonal rains in the tropics. Scientific evidence shows that some fly over the Himalaya and cross the Indian Ocean before arriving to breed in East Africa. It is believed that their progeny then move into southern Africa, with individuals from subsequent generations making the return journey.

LOOK-ALIKES

Darters
(*pages 149–150 & 180–182*)

Female skimmers
(*pages 136–138 & 172*)

WHERE TO LOOK	OBSERVATION TIPS
Perhaps most likely to turn up near the south coast during hot, southerly or south-easterly winds. Might be attracted to light (*e.g.* a lighthouse or moth trap).	Look for its high-flying, gliding flight action with abdomen angled down. If you find one perched, try to get a photograph or video sequence.

The broad, triangular hindwing enables sustained flight in both sexes

The tapering, cylindrical abdomen has an irregular, though quite intricate, dark marking down the top

Introduced exotic species

Thirteen species – five damselflies and eight dragonflies – have been recorded in Britain only as a result of accidental introductions, either as eggs or larvae in imported aquatic plants. These species have been included in this book to increase awareness that further potential confusion species could be encountered.

Most of the exotic species recorded in Britain have been found in the greenhouses of importers of tropical pondweeds. However, at least one, Common Bluetail *Ischnura senegalensis*, has been discovered at a garden pond, to which it was probably moved with recently imported pondweed. Although unlikely, there is a possibility that these, or indeed other, Dragonfly species could be encountered, particularly near commercial aquatic greenhouses. As the species concerned have originated in hot climates, it is even more unlikely that successful establishment could occur in the wild. Those that have been recorded are listed in the table below, with details of their regions of origin and the British or Irish species with which confusion is most likely. Photographs of some examples are shown opposite. The English names used are those generally adopted in field guides to the areas concerned.

The incidence of exotic species will be related to the origins and extent of trade in pondweeds and water-lilies, and the success of phytosanitory procedures at the points of export and import. Much recent trade appears to centre on Singapore in particular, so common Asian species such as Common Bluetail and Oriental Scarlet are the more likely to be found. Such species may be very similar to some that occur naturally in Britain or Ireland, in these examples Blue-tailed Damselfly and Scarlet Darter, respectively. These similarities are confounded because emergents or tenerals are most likely to be found. Importing wholesalers and plant nurseries are the most likely places to find larvae or emergent adults, although some may emerge from domestic indoor aquaria after sale. The latter have produced several records, so clearly any unusual species found indoors or around houses or garden ponds needs careful checking. A representative from the British Dragonfly Society will try to identify images emailed to **BDS@british-dragonflies.org.uk**.

	SPECIES	ORIGIN	Look-alikes
Damselflies	Variable Dancer *Argia fumipennis*	North America	Blue damselflies
	Ornate Coraltail *Ceriagrion cerinorubellum*	South-East Asia	None in Britain/Ireland
	Orange Bluet *Enallagma signatum*	USA	Blue-tailed damselflies
	Fragile Forktail *Ischnura posita*	North America	Blue damselflies
	Common Bluetail *Ischnura senegalensis*	Asia and Africa	Blue-tailed damselflies
Dragonflies	Green Emperor *Anax gibbosulus*	Australasia	Emperor dragonflies
	Pale-spotted Emperor *Anax guttatus*	Orient (Asia)	Emperor dragonflies
	Oriental Scarlet *Crocothemis servilia*	Asia	Scarlet Darter
	Eastern Pondhawk *Erythemis simplicicollis*	N. & C. America	Skimmers
	Slender Skimmer *Orthetrum sabina*	N. Africa to Australasia	Skimmers, clubtails
	Common Redbolt *Rhodothemis rufa*	South-East Asia	Darters
	Saddlebag Glider *Tramea transmarina*	South-East Asia	Chasers
	A basker *Urothemis bisignata*	Philippines	Darters

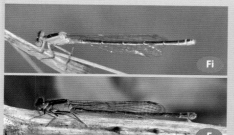

Fi

F

Common Bluetail

M

Ornate Coraltail

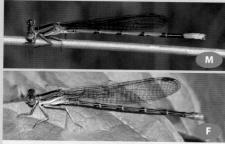

M

F

Variable Dancer

M

F

Fragile Forktail

M

Pale-spotted Emperor

F

Slender Skimmer

Oriental Scarlet

F

Oriental Scarlet

M

Common Redbolt

Identification of larvae and exuviae

An introduction to identification

Most people who become interested in dragonflies do so through a fascination with the adult forms, rather than larvae. For this reason, the main focus of this book is on adult identification. However, there are good reasons for taking an interest in larvae and exuviae, since their presence and numbers at a water body give the best indication of breeding populations and hence the conservation value of a site.

Larval identification requires different skills and equipment to both catch and identify specimens. They can be caught with a pond net or sieve, or by studying pondweed and organic debris. The catch is best examined in a white tray filled with a little water. Simply wait until something moves – damselflies wriggle and dragonflies walk or shoot forward by squirting water out of their anus. They can then be transferred into a smaller, shallow container to allow closer scrutiny.

Finding a newly emerged adult Dragonfly often leads to the discovery of an exuvia and provides ultimate proof that a species has bred successfully at a site. Pre-flight emergents (adults before their maiden flight) can help you to identify the exuviae near them, provided that they have coloured up a little; newly emerged specimens lack the characteristic colours of a mature adult and any patterning is very subdued. You may disturb an emergent Dragonfly and prompt its maiden flight: this will be very weak, typically only a short distance, and its wings will be very reflective. Avoid handling emergents, however, because they are still soft and easily damaged. As it is the cast skin of the final stage larva, an exuvia has all the features needed for correct identification, but is very fragile and easily damaged or blown away. Exuviae can, however, be collected for examination indoors and stored in labelled containers.

Recording the species and numbers of Dragonflies emerging from a site is critical to evaluating its viability and conservation importance – and can be a most satisfying activity. Although counts of mature adults are often used to evaluate a site, the numbers of exuviae in fact provide a more robust means of monitoring productivity. Hawkers emerge at night, but most species do so early in the day – morning is therefore the best time to check the margins of sites. Close-focus binoculars are a surprisingly useful aid to finding exuviae. It is important to realise that exuviae will be only about half the length of the adults that leave them. Most exuviae will be found on emergent or marginal vegetation within a few centimetres of the water's surface. However, it is not unusual for them to be found on artefacts, and sometimes even some metres away from the margin.

Identification tips

This section takes a pragmatic view of the identification of larvae or exuviae and concentrates on what can be achieved using the naked eye, a hand lens or binoculars (look through the 'wrong' end, with the specimen held close to the eyepiece). Identification of the 'type' of Dragonfly (see *pages 196–201*) is not too difficult, and it is even possible to identify some species using certain key features. However, specific identification, especially of damselflies, often requires a critical examination of hairs on the labium or caudal lamellae, for which a good ×10 hand lens or, ideally, a microscope is needed. Debris often conceals details, so wash larvae or gently brush exuviae to improve viewing. The larvae of blue damselflies and some darters are notoriously difficult to identify. Where unusual species are suspected try and confirm the identification by looking for adults. The keys included in several of the reference books suggested for further reading (*page 218*) provide additional information on microscopic characters, while Steve Cham's field guide provides the most detailed descriptions.

How to use this guide

The tables on the following pages provide a summary of the habitat preferences and biology of the early life-stages of all Dragonfly species that are known or are suspected to have bred successfully in Britain or Ireland. These are followed by annotated illustrations that highlight the key characteristics of each of seven 'types' of damselfly and three broad groups of dragonfly. Diagnostic details are given that should enable the exuviae or final-stage larvae of most species to be identified. Final-stage larvae have wing sheaths that extend over the abdomen as far as the fourth or fifth segment (S4 or S5) – it is easiest to count backwards from the final, tenth segment. As with the descriptions of adult Dragonflies, key identification features are highlighted in red text. The sizes given for each species indicate the range in length of full-grown larvae (or, if unknown, exuviae), which are best found in the weeks prior to typical emergence times.

The biology of eggs and larvae

The table presented here provides information, where known, on the timings and duration of the early life stages of all the Dragonfly species that have, or are suspected to have, bred in Britain or Ireland. They also indicate the micro-habitat preferences of the larvae, which generally relate to the species' favoured broad habitat types summarised in the tables on *pages 30–31*. The peak emergence period is the optimum time to search for exuviae. The gaps in the table indicate how relatively little is known about the early life stages of Dragonflies, and provide a useful pointer to where valuable scientific contributions can still be made.

DAMSELFLIES	Eggs hatch ...	Larval micro-habitat	Adult emergence After ...	Peak	*Page*
Beautiful Demoiselle	after 2 weeks	Vegetation, debris and roots	2 years	May–June	*203*
Banded Demoiselle	after 2 weeks	Vegetation, debris and roots	1–2 years	May–June	*203*
Emerald Damselfly	next spring	Vegetation	2–3 months	July	*203*
Scarce Emerald Damselfly	next spring	Vegetation	2 months	June	*203*
Southern Emerald Damselfly	next spring	Vegetation	2–3 months	?	*202*
Willow Emerald Damselfly	next spring	Vegetation	2–3 months	May–June	*202*
Large Red Damselfly	after 2–3 weeks	Pondweeds and debris	1–3, usually 2, years	May	*203*
Small Red Damselfly	after 1 month	Bog-mosses, other plants and debris	2 years	June–July	*203*
White-legged Damselfly		Debris around emergent plant roots	2 years	May–June	*205*
Southern Damselfly	after 4–6 weeks	Vegetation	2 years	June	*205*
Northern Damselfly	after a few weeks	Vegetation	2 years	June	*205*
Irish Damselfly	after a few weeks?		2 years?	(June)	*205*
Dainty Damselfly	after 6 weeks	Pondweeds	1 year	June	*205*
Azure Damselfly	after a few weeks	Vegetation	1 (south) or 2 years	May	*205*
Variable Damselfly	after 1 month	Vegetation	probably 1 year	May	*205*
Common Blue Damselfly		Vegetation, including drifting mats far from margins	1 (south), 2 or more years	May–June	*205*
Blue-tailed Damselfly		Vegetation	1 (south) or 2 years	May–June	*205*
Scarce Blue-tailed Damselfly	after 2–3 weeks	Silt	1, perhaps 2, years	June	*205*
Red-eyed Damselfly		Vegetation	1, or usually 2, years	May	*203*
Small Red-eyed Damselfly		Pondweeds	1 year	July	*203*

DRAGONFLIES	Eggs hatch ...	Larval micro-habitat	Adult emergence		Page
			After ...	Peak	*Page*
Hairy Dragonfly	after 3–4 weeks	Rotting plant debris	(1)–3 years	May	206
Azure Hawker	next year	Detritus at bottom of bog pools	3–4 years	July	206
Common Hawker	next year	Bog-mosses; moves to bottom as matures	3 or more years	July	207
Migrant Hawker	next spring	Debris below emergent plants	a few months	August	206
Southern Migrant Hawker	after winter inundation	Probably debris below emergent plants	1–2 years	June	206
Southern Hawker	next spring	Vegetation and organic debris	1–3 years	July	207
Brown Hawker	next spring	Plants/detritus near the bottom	2–4 years	July	207
Norfolk Hawker	after 3–4 weeks	Vegetation and rotting plant debris	2 years	May–June	206
Emperor Dragonfly	after about 3 weeks	Vegetation	1, usually 2, years	June	207
Lesser Emperor	after some weeks	Vegetation or bottom debris	1–2 years	August	207
Golden-ringed Dragonfly		Bottom debris and sediment	2–5 years	June	211
Common Club-tail		Fine sediment	3–5 years	May	211
Downy Emerald	after 2–3 weeks	Leaf-litter, coarse organic debris		May	208
Brilliant Emerald	soon after being laid or next spring?	Decaying debris, often under trees		June	208
Northern Emerald	soon after being laid or next spring?	Shallow water in decaying bog-mosses	2(+?) years	June–July	208
Four-spotted Chaser		Decaying plant debris	2(+?) years	May–June	209
Broad-bodied Chaser	after 2–3 weeks	Sediment, often covered in silt	1–3 years	May	209
Scarce Chaser		Mud below tall, emergent plants	2 years	May	209
Black-tailed Skimmer	after 5–6 weeks	Bottom debris	2–3 years	June	209
Keeled Skimmer		Bottom debris	2 years	June–July	208
White-faced Darter		Half-submerged bog-mosses	1–3, usually 2, years	May–June	211
Black Darter	next spring	Bog-mosses or other vetetation, or mud	2 months	July–August	211
Common Darter	after a few days, or next spring	Mud and aquatic vegetation	1 year	July–August	210
Ruddy Darter	after a few days if wet, or next spring	Roots of emergent plants in shallow water	usually 1 year	July	210
Red-veined Darter	soon after being laid	Mud and aquatic vegetation	2–3 months or 1 year	June, Sept.	210
Yellow-winged Darter	next spring	vegetation or at bottom	a few months	(July)	211

KEY FEATURES OF LARVAE & EXUVIAE

The terms used in this book for the various parts of damselfly and dragonfly larvae are shown on the following illustrations.

Key features of a damselfly larva	*Large Red Damselfly ×4*

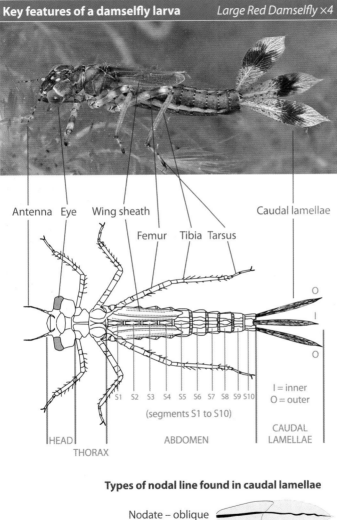

Antenna Eye Wing sheath Caudal lamellae

Femur Tibia Tarsus

S1 S2 S3 S4 S5 S6 S7 S8 S9 S10
(segments S1 to S10)

I = inner
O = outer

HEAD
THORAX ABDOMEN CAUDAL LAMELLAE

To identify damselfly species look at:

Overall shape and size

Head shape and spotting

Antennae

Labium shape and characteristics

Caudal lamellae shape and patterning

Dark banding on legs

Caudal lamellae come in a variety of shapes, some of which are diagnostic, although most details cannot be seen easily without a microscope. Lamellae from exuviae are best examined after soaking in water with a little vinegar or washing-up liquid.

Remove a lamella, ideally the middle one, with a pair of tweezers and place it on a glass slide for examination.

Lamellae that show a line between the thickened basal half and the thinner, terminal section are termed **nodate**; this nodal line may be perpendicular to the edge or oblique. If this line is indistinct, the term used is **sub-nodate**.

Some lamellae are marked with irregular spots, bands or other shapes.

Note that some or all lamellae may be lost, without apparent harm to the individual.

Types of nodal line found in caudal lamellae

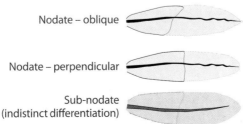

Nodate – oblique

Nodate – perpendicular

Sub-nodate
(indistinct differentiation)

Key features of a dragonfly larva

Other dragonfly groups are more 'spider-like' in shape from above
(see page 200–201)

To identify dragonfly species look at:

Overall shape and size
Antennae
Head shape
Labium shape
Relative leg length
Abdominal spines (top and side)
Patterning on abdomen and legs
Anal appendages

Antenna   Eye   Wing sheath   Spines (side)

Femur Tibia Tarsus   Anal appendages

S1 S2 S3 S4   S5   S6   S7   S8   S9 S10

(segments S1 to S10)

HEAD   THORAX   ABDOMEN   ANAL APPENDAGES

The **anal appendages** of dragonflies comprise a central epiproct flanked by a pair of paraprocts and above them two cerci.

The relative lengths of the cerci and the shape of the tip of the epiproct are useful in the identifcation of some species.

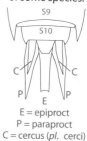

S9
S10
C   C
P   E   P
E = epiproct
P = paraproct
C = cercus (*pl.* cerci)

Labium shapes of damselfly and dragonfly larvae

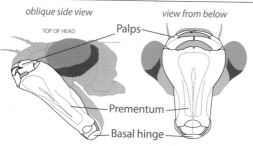

oblique side view

TOP OF HEAD

Palps

view from below

Prementum

Basal hinge

Flat with pincers
Damselflies, hawkers and Common Club-tail

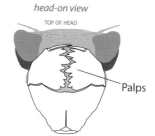

head-on view

TOP OF HEAD

Palps

'Spoon-shaped' with 'teeth'
All other dragonflies

THE TYPES OF DRAGONFLY LARVAE

The following silhouettes, shown at life-size, along with the photographs (×2 life-size) aim to help identify a **full-grown larva** (*i.e.* where the wing sheaths extend beyond the third abdominal segment (S3)) **or exuvia** to species, or, in some cases, a group of species. Once assigned to a group, reference to the relevant Larval Identification Chart on *pages 202–211* should allow specific identification. It should be noted, however, that the specific identification of some specimens may not prove possible, and since larval colour is highly variable, this is not usually a reliable feature.

Length is measured from the front of the head (excluding the projection of the antennae) **to the tip of the anal appendages** (caudal lamellae or paraprocts).

DAMSELFLIES (Zygoptera)

All damselflies have up to three caudal lamellae and a flattened labium.

DEMOISELLES Genus: *Calopteryx*
(2 species)

Banded Demoiselle

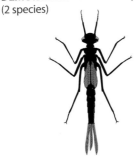

Found in running water. Stiff and stick-like, with small head, dark stripes along the body and across the long, thin, spidery legs. Long antennae, with basal segment half the total length (all other damselflies have the segments of roughly equal length). Labium has a narrow base, and the middle lobe a deep cleft. The lamellae are thin, the outermost being longer and triangular in cross-section.

EMERALD DAMSELFLIES Genus: *Lestes*
(4 species)

Scarce Emerald Damselfly

Slender, long-legged larvae with a relatively broad head; a pale line down the back; a ladle-shaped labium with stalk-like prementum (or labium shorter and triangular in one species), with a slit-like cleft at the tip and antler-shaped palps; and two or usually three dark bands across rigid lamellae.

BLUE-TAILED DAMSELFLIES
Genus: *Ischnura* (2 species)
Scarce Blue-tailed Damselfly

Note: the blue damselflies are difficult to identify without microscopic examination.

BLUE DAMSELFLIES
Genera: *Coenagrion, Enallagma* (6 species)
Azure Damselfly

Similar to blue damselflies, but with no speckling on the head; the caudal lamellae are long and narrow (length = 4× width), with pointed tip and stout hairs to the mid-point on one edge, and are obliquely sub-nodate.

Caudal lamellae are nodate, having a distinct junction between the thickened base and thinner terminal half; the abdomen is marked with a pale mid-line; the rear of the head and the wing buds have fine, dark speckling (but absent in Common Blue); and antennae have 6 or 7 segments.

RED-EYED DAMSELFLIES
Genus: *Erythromma* (2 species)
Red-eyed Damselfly

Caudal lamellae are bluntly-rounded and nodate with three bands or blotches on the terminal half. The labium is relatively long and narrow.

RED DAMSELFLIES
Genera: *Pyrrhosoma, Ceriagrion* (2 species)
Large Red Damselfly

Squat; abdomen relatively short; brownish in colour; head broad, hardly tapering behind eyes and with rectangular corners to the rear.

WHITE-LEGGED DAMSELFLY
Genus: *Platycnemis* (1 species)
White-legged Damselfly

Found in muddy rivers. Long, hairy legs. Distinctive caudal lamellae, with long, narrow pointed tip, dark blotches and long hairs of variable length on the margins.

DRAGONFLIES (Anisoptera)

Dragonfly larvae fall into three broad groups: Hawkers, Common Club-tail and 'others'. The first two groups have a flat labium, while the 'others' (Golden-ringed Dragonfly, emeralds, chasers, skimmers and darters) have a 'spoon'-shaped labium that bulges under the front of the head, creating the impression of a 'beak' when viewed from the front (see *page 195*). These 'other' species are mostly squat and spider-like in appearance.

HAWKERS Genera: *Brachytron, Aeshna, Anax* (9 species)

| Hairy Dragonfly | Brown Hawker | Emperor Dragonfly |

Hawker larvae are at least 30mm in length; have long, thin torpedo-shaped bodies; a flat labium with pincers at the front; and 7 (occasionally 6) antennal segments. They fall into three groups: **Hairy Dragonfly**, the *Aeshna* **hawkers** and the **emperor dragonflies**, which can be separated by the relative size and shape of their eyes, and the shape of their heads (see *opposite*).

Once the specimen has been assigned to a group, a combination of characteristics is then required to identify to species level.

These are (see *opposite*):

- **The height–to–width ratio of the labium;**
- **The presence/absence and relative length of abdominal lateral spines**; and
- **The shape of the epiproct.**

These are summarised *opposite* and detailed in the relevant section of the chart (see *pages 206–207*).

Hairy Dragonfly (*Brachytron*)

Hawkers (*Aeshna*)

Emperor dragonflies (*Anax*)

Hawker head characteristics

Hawkers can be grouped by the relative size and shape of their eyes when viewed from above:
compare 'eye length' (**E**) with 'head length' (**H**)

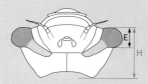

Hairy Dragonfly:
small eyes
(**E** much less than half the length of **H**);
head tapers strongly to the rear.

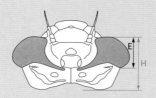

Aeshna hawkers:
large eyes
(**E** about half as long as **H**);
head pentagonal from above with angled rear edge.

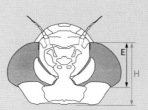

Emperor dragonflies:
very large eyes
(**E** much greater than half the length of **H**);
head rather rounded from above with straight rear edge.

Hawker labium ratio

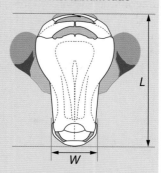

Hawkers have 3 basic labium shapes. This can be assessed by calculating the ratio of the total length of the mask (including the basal hinge, prementum and palps) divided by the basal width of the prementum (*i.e.* $L \div W$).

Hawker lateral spine length

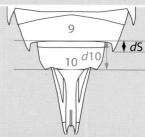

Hawkers fall into 3 groups based on the relative length of the spine on S9 (black arrow **dS**). This can be expressed as a proportion of the length of S10 (orange arrow **d10**).

Epiproct tip shape

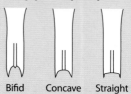

Bifid   Concave   Straight

The shape of the tip of the epiproct can aid identification.

CLUB-TAILS Genus: *Gomphus* (1 species)

Common Club-tail

Squat, flattened and spider-like, with flat labium and small, heart-shaped head from above; antennae with only 4 broad segments, the third antennal segment being as long as the other three combined (all other dragonflies have 7, occasionally 6, segments); occurs in muddy rivers, where silt clings to its abundant body hairs; front and middle tarsi have 2 (not 3) segments.

GOLDEN-RINGED DRAGONFLY Genus: *Cordulegaster* (1 species)

Golden-ringed Dragonfly

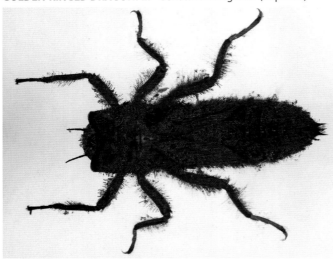

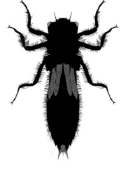

Lives in gravel-bottomed running waters, where silt and debris cling to its abundant body hairs; short legs; small, bulging eyes; squat body; spoon-shaped labium with characteristic irregular, jagged 'teeth'; lateral spines on S8 and S9.

EMERALDS Genera: *Cordulia, Somatochlora* (3 species)

Downy Emerald

Labium spoon-shaped with deep serrations on the palps; head tapering markedly to the rear; very long legs, hind legs reaching well beyond the tip of the abdomen; cerci more than half the length of the paraprocts; sides of abdomen lack long spines.

CHASERS and SKIMMERS Genera: *Libellula, Orthetrum* (5 species)

Four-spotted Chaser

Labium spoon-shaped; head rectangular from above; extended hind legs reach beyond (chasers) or just reach (skimmers) the tip of the abdomen; dorsal spine on S8 present (chasers); or absent (skimmers); cerci less than half the length of the paraprocts.

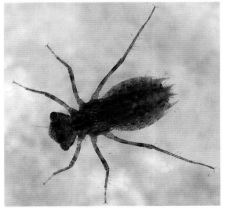

DARTERS Genera: *Sympetrum, Leucorrhinia* (6 species)

Common Darter

Labium spoon-shaped; head tapering markedly to the rear from above; legs long and slender, extending well beyond the tip of the abdomen; cerci less than half the length of the paraprocts. The relative lengths of lateral spines on S8 and S9 are important in identification (see *page 210*). A difficult group to identify.

DEMOISELLES, EMERALD, RED and RED-EYED DAMSELFLIES (Larvae 1·5× life-size)

DEMOISELLES Genus: *Calopteryx*

ANTENNAE:
Basal segment half length of antenna

LABIUM: Narrow base; middle lobe with deep cleft

LARVA COLOUR:
Beautiful usually well patterned

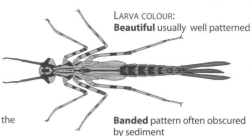

Beautiful Demoiselle has a thorn-like protruberance on the side of the head behind the eye (absent on Banded)

Banded pattern often obscured by sediment

EMERALD DAMSELFLIES Genus: *Lestes*

ANTENNAE:
7 segments

All damselflies except demoiselles have short basal segments

LABIUM: Ladle-shaped; very narrow base

HEAD: Broad compared to slender body

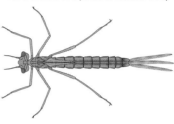

Southern Emerald Damselfly (*page 68*) has labium and lamellae very similar to Scarce Emerald Damselfly.

Willow Emerald Damselfly (*page 70*) is similar in shape to Emerald Damselfly but the labium is shorter and more triangular in shape.

THORAX COLOUR: Translucent green

RED DAMSELFLIES Genera: *Pyrrhosoma, Ceriagrion*

HEAD:
Broad; rectangular rear edge

ANTENNAE:
7 segments

LABIUM: **Typical**
(*see page 204*)

LARVA: Chunky, with relatively long wing sheaths (to S5/6) and short abdomen

LARVA COLOUR:
Reddish- brown

LARVA: Chunky, with relatively short abdomen; large eyes

LARVA COLOUR:
Usually brownish or olivaceous

RED-EYED DAMSELFLIES Genus: *Erythromma*

ANTENNAE:
6 segments

HEAD: Relatively small, tapering at rear

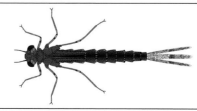

ANTENNAE:
7 segments

LABIUM: **Typical**
(*see page 204*)

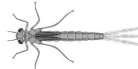

CAUDAL LAMELLAE 4× life-size			HABITAT	LARVAL LENGTH (range)
Beautiful Demoiselle (*page 60*) outer central	Outer lamellae narrow and triangular in section	6–10mm; dark with usually one pale band across middle	Running water with stony and gravelly bottom	30–35mm
Banded Demoiselle (*page 62*) outer central		9–14mm; dark with usually 2 pale bands across middle	Running water with muddy bottom	30–40mm
Emerald Damselfly (*page 64*) outer central	Usually 3 dark bands	9–10mm; large; outer lamellae with parallel sides	Well vegetated, shallow standing waters, often with rushes	26·5–34·5mm
Scarce Emerald Damselfly (*p. 66*) outer central	Usually 2 dark bands on central lamella	9–10mm; large; outer lamellae curved and tapering	Well vegetated, coastal ditches and shallow ponds with fluctuating water levels	29–32mm
Large Red Damselfly (*page 72*)	Broad near tip, tapering abruptly Fine hairs on terminal half	4–6mm; blotches on terminal half, usually forming 'X'-shape	A wide range of standing or slow-flowing waters	19–22·5mm
Small Red Damselfly (*page 74*) Sub-nodate	Short, broad, tapering abruptly No hairs on terminal half	4–5mm; obscure, irregular blotches on edge; veins prominent	Acidic bogs, streams and ponds	16–17mm
Red-eyed Damselfly (*page 96*) Nodate, perpendicular to margin	Bluntly rounded Coarse hairs on basal margins, not beyond node	8–9mm; 3 broad, dark bands across terminal half	Floating leaves, especially water-lilies, and algae	29–32mm
Small Red-eyed Damselfly (*p. 98*) Nodate, oblique to margin	Rounded tip Coarse hairs to node only	5–6mm; usually 3 dark blotches on terminal half	Floating algae, hornwort and water-milfoil	19–21mm

WHITE-LEGGED, BLUE-TAILED and 'BLUE' DAMSELFLIES   (Larvae 1·5× life-size)

WHITE-LEGGED DAMSELFLY Genus: *Platycnemis*

ANTENNAE:
7 segments

LABIUM: **Typical Damselfly structure**

HEAD SHAPE:
Rectangular rear edge

LEGS:
Long and hairy

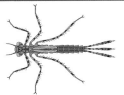

BLUE-TAILED DAMSELFLIES Genus: *Ischnura*

ANTENNAE:
7 segments

LABIUM: **Typical**

LEGS:
Dark band near tip of femora

LEGS:
No, or indistinct, dark bands on femora

'BLUE' DAMSELFLIES Genera: *Coenagrion, Enallagma*

ANTENNAE:
6 segments

LABIUM: **Typical**

HEAD SPECKLING:
None

LARVA COLOUR:
Usually greenish

Dainty Damselfly is similar to other 'blue' damselflies but has 7 antennal segments and no speckling on head

ANTENNAE:
7 segments

LABIUM: **Typical**, relatively short and wide

HEAD SPECKLING:
Distinct speckling on rear

NOTE: Many **Azure** and **Variable Damselflies** cannot be reliably identified as larvae or exuviae due to an overlap in characteristics (see Cham 2012)

ANTENNAE:
7 segments

LABIUM: **Typical**

HEAD SPECKLING:
Distinct speckling on rear

HEAD SPECKLING:
Speckling on rear irregular, sparse at edge

HEAD SPECKLING:
Speckling on rear evenly distributed

ANTENNAE:
6 segments

LABIUM: **Typical**

CAUDAL LAMELLAE 4× life-size			HABITAT	LARVAL LENGTH (range)
White-legged Damselfly (*page 76*)	Long narrow point at tip	6–7 mm; hairs long, of varying length, often coated in debris	Slow-flowing muddy rivers	19–22 mm
Blue-tailed Damselfly (*page 92*)	Long and narrow, length 4× width, tip pointed	5–6 mm; stout hairs to mid-point on one edge and one third along the other	Widespread in lowlands	18–21 mm
Scarce Blue-tailed Damselfly (*p. 94*)	Sub-nodate, oblique to margin but indistinct	4–6 mm	Small streams, flushes and ponds with sparse vegetation	15–18 mm
Common Blue Damselfly (*page 90*)	>3 mm long, length 3× width Sub-nodate	6 mm; stout hairs to mid-point 1–3 narrow, dark bands	Lakes, reservoirs, large ponds and canals	20–26·5 mm
Dainty Damselfly (*page 84*)	>3 mm long, length 4× width, Sub-nodate	5–7 mm; bluntly pointed	Ponds and ditches	16–21 mm
Southern Damselfly (*page 78*)	Short (<4 mm), boat-shaped, pointed tip NOT nodate	3–4 mm; dense, long hairs after mid-point	Acidic runnels, chalk streams, ditches and occasionally fens	15–17 mm
Azure Damselfly (*page 86*)	Nodate, may be perpendicular to margin	5–6 mm; evenly tapered at the base	Ponds and ditches	22–25·75 mm
Variable Damselfly (*page 88*)	Nodate, may be oblique to margin	5–6 mm; broader and tending to bulge at their base	Well-vegetated ditches and standing waters	20–25·25 mm
Irish Damselfly (*page 82*)	Nodate, perpendicular to margin	6 mm; nodal line unpigmented	Shallow, clear ponds and lakes with abundant vegetation	21–23 mm
Northern Damselfly (*page 80*)		5–6 mm; dark nodal line and dark nodal constriction	Acidic pools and loch margins	19–21 mm

LARVAL IDENTIFICATION CHART: DRAGONFLIES 1

HAIRY DRAGONFLY, HAWKERS and EMPERORS (Larvae life-size)

Species	Hairy Dragonfly (page 100)	Migrant Hawker (page 106)	Southern Migrant Hawker (page 108)	Azure Hawker (page 102)	Norfolk Hawker (page 114)
Habitat	Decaying stems below emergent plants	Debris below emergent plants, mainly in still water		Decaying bog-moss in small bog pools	Ditches with abundant Water-soldier and clear water
Antennae	7 segments	7 segments		6 segments	7 segments
Head and eyes (see page 199) HEAD SHAPE FROM ABOVE: EYES:	Strong taper to rear Small, **E** less than ½ **H**	Pentagonal, hind margin angled Large, **E** about ½ **H**			
Labium ratio $L \div W$ (see below)	≈ 3	> 3	≈ 2·8	≈ 3	≈ 3
Larva (bars show the maximum and minimum lengths of full-grown larvae)	35–40 mm ANAL APPENDAGES: length shorter than S9+S10 length	30–38 mm	30–39 mm	34–39 mm	38–44 mm
Abdominal spines (side) S5	tiny	×		×	×
S6	present	present		× / tiny	present
S9 ratio (= $d10 \div dS$) (see p. 199)	ratio ≈ 0·3 Also tiny spine on top of S9	ratio ≈ 1·0	ratio <0.65	ratio ≈ 0·3	ratio ≈ 0·6 ± 0·1
Epiproct tip shape (see page 199)		Straight	Straight	Bifid	Straight
Cerci/paraproct ratio (see page 195)		< 1/2		< 1/2	2/3

Labium ratios

Details of how to calculate the labium ratio are given on page 199. The Illustrations opposite show the shape of the labia for each of the three types found in the British hawkers. With practise the extremes can be detected with the naked eye.

 ≈ 4

 ≈ 3

 ≈ 2·5

Southern Hawker (page 110)	Common Hawker (page 104)	Brown Hawker (page 112)	Lesser Emperor (page 118)	Emperor Drag. (page 116)
Pondweeds and coarse debris in ponds	Mainly acidic ponds and bog pools	Coarse debris in mainly still waters	Pondweeds and debris in still waters	Pondweeds in various still and flowing waters
Typical Dragonfly antenna (7 segments)			**Typical** 7 segments	
Pentagonal, hind margin angled Large, **E** about ½ **H**			Rounded with straight hind margin Very large, **E** much more than ½ **H**	
≈ 4	≈ 2·5	≈ 3	≈ 4	
		A pale stripe on each side of thorax and head		
38–48 mm	40–51 mm	40–46 mm	48–54 mm	45–56 mm
✕	✕	✕ / tiny	✕	
present	✕ / tiny	present	✕	
ratio ≈ 0·6 ± 0·1	ratio ≈ 0·3	ratio ≈ 0·6 ± 0·1	ratio ≈ 0·6 ± 0·1	
Concave	Bifid	Bifid	**Lesser Emperor** Base projection of male epiproct half as long as broad; one-third length of cerci	**Emperor Drag.** Base projection of male epiproct as long as broad; half length of cerci
< 1/2	< 1/2	< 1/2		

LARVAL IDENTIFICATION CHART: DRAGONFLIES 2

EMERALDS, SKIMMERS and CHASERS   (Larvae life-size)

Species	Downy Emerald (page 124)	Brilliant Emerald (page 126)	Northern Emerald (page 128)	Keeled Skimmer (page 138)
Habitat	Leaf litter in ponds in or near woodland	Decaying plant debris under trees or in peaty lochans	Decaying bog-moss in boggy pools and lochs	Bottom debris in acidic bogs, streams and seepages
Appearance	LEGS: Well-patterned THORAX: Stripy sides ABDOMEN: Rear end truncated; rows of pale yellow dorso-lateral spots	ABDOMEN: Rather cigar-shaped; dark spots along the side of each segment	THORAX AND ABDOMEN: Very hairy; uniformly-coloured	ABDOMEN: Unmarked
Head	HEAD SHAPE FROM ABOVE: Tapered LABIUM: 8–10 regular broad deep 'teeth'			HEAD SHAPE FROM ABOVE: Rectangular
Larva (bars show maximum and minimum lengths of full-grown larvae)	22·5–25 mm	24–25 mm	17–22·5 mm	17–23 mm
Abdominal spines **Top**	S4–S8 relatively small; S9 tiny	S4–S8 prominent; S9 prominent	Spines absent; stout setae present	S4–S7 very short; S8–S9 absent
Side S8–S9	Short	Short	Absent	Very short
Cerci length (see page 195)	> half length of paraprocts			< half length of paraprocts

Larva of **Downy Emerald** (at 2× life-size), showing the spoon-shaped labium (*right*); the diagnostic stripy sides to the thorax and the prominent dorsal spines.

Black-tailed Skimmer *(page 136)*	Four-spotted Chaser *(page 130)*	Broad-bodied Chaser *(page 132)*	Scarce Chaser *(page 134)*
Muddy debris in lowland waters with bare margins	Fine decaying plant debris in well-vegetated still waters	Muddy debris in small, still, often new waters	Muddy debris below emergent plants, mainly in slow-flowing waters
ABDOMEN: 4 rows of dark spots on top	ABDOMEN: Rear end not truncated; unmarked, sepia coloured	ABDOMEN: Tends to be broader and rear end more truncated; dark stripes on top; often muddy	ABDOMEN: Unmarked
HEAD SHAPE FROM ABOVE: Rectangular	EYES: Point outwards, tips below top of head LABIUM: Protruding strongly; 8–9 very shallow 'teeth'	EYES: Raised markedly above rest of head LABIUM: 7–8 rounded, fairly deep 'teeth'	HEAD SHAPE FROM ABOVE: More or less rectangular
23–25 mm	22–26 mm	22·5–25 mm	22–25 mm
S3–S6 very short; S7–S9 absent	S4–S8 short; S9 absent	S4–S8 short; S9 absent	S4–S9 long and prominent
Short	Short	Short	Short
< half length of paraprocts	> half length of paraprocts	< half length of paraprocts	< half length of paraprocts

Larva of **Four-spotted Chaser** (at 2× life-size), showing the spoon-shaped labium (*left*); the eyes level with the top of the head and the prominent dorsal spines.

DARTERS (Larvae life-size)

Species	Common Darter (page 144)	Ruddy Darter (page 146)	Red-veined Darter (page 148)
Habitat	Debris and pondweeds in a wide range of waters	Shallow still waters choked with emergent vegetation	Shallow standing waters, fresh or brackish
Larva (bars show maximum and minimum lengths of full-grown larvae)	15·5–18 mm	15–17 mm	15–18 mm
ABDOMEN Top spines	S4 sometimes tiny; S4–S8 slightly curved	S4 tiny, sometimes absent; S5–S8 slightly curved	Spines absent; stout bristles present
Side spines on S8 & S9 (See box below for explanation)	>0·4 mm >1·0 mm; S9 usually incurved; S9 spine > *d*; and S8 >0·4 mm; S9 >1·0 mm	<0·4 mm <1·1 mm; S9 usually incurved; S9 spine < *d*; and S8 <0·4 mm; S9 <1·1 mm	S8 tiny; S9 very short; prominent dark stripes on upperside of abdomen

Lateral spine length measurement in darters

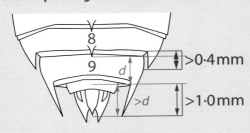

>0·4 mm

>1·0 mm

The length of the spine on S9 is a key factor in the identification of darter larvae.

The differences between the species are shown in the table above, where the **length of S9** is shown as *d* (orange arrow) and the **length of the spine on S9** is shown as a **red arrow**.

It is important to note that it is not always possible to differentiate Common and Ruddy Darters on the basis of the relative spine length, but the actual lengths of the side spines on S8 and S9 (measurements in **blue text**) may help.

Yellow-winged Darter (page 150)	Black Darter (page 142)	White-faced Darter (page 140)
Shallow, often acidic, waters with abundant tall vegetation	Bog-mosses in acidic upland and lowland bog pools	Bog-mosses in acidic lowland bog pools
14–17 mm	13–16 mm	18–20 mm
S4 absent; S5–S8 present	S4 absent; S5–S7 present; S8 tiny, sometimes absent	S4–S6 short; S7 tiny/absent; S8–9 absent
S9 spine about 0·5 d	S9 spine about 0·33 d	three dark stripes on underside of abdomen S8 & S9 spines present

COMMON CLUB-TAIL and GOLDEN-RINGED DRAGONFLY   (Larvae life-size)

Species	Common Club-tail (page 122)	Golden-ringed Dragonfly (page 120)
Larva (bars show maximum and minimum lengths of full-grown larvae)	Very hairy; mud clings to abdomen hairs LEGS: Spider-like, front and middle tarsi with 2 segments 27–30 mm	Very hairy; debris clings to abdomen hairs LEGS: Tarsi with 3 segments 35–42 mm
Antenna	AS3 4 broad segments, AS3 long	Typical
Habitat	Muddy sediment	Gravel or other bottom sediment

A signficant number of the Dragonfly species in Britain and Ireland are rare or highly localized. As a consequence, many are of conservation concern and those that are particularly at risk are afforded legislative protection, including through requirements to conserve their habitats. This legislation is informed by evaluations of the status of all species. The legal measures and the means by which species are assessed are explained in this section and summarised in the table below. This information is highlighted in the red 'Legal Status' boxes in the main species accounts (see explanation on *page 59*) and an explanation of the codes used is given in the following sections.

Red Lists

The International Union for Conservation of Nature (IUCN) has established global criteria for assessing the conservation status of species. These criteria include the rate of decline, population size, area of geographic distribution, and degree of population and distribution fragmentation. Based on an analysis of these criteria, species are assigned to one of the following categories: **Extinct (EX), Endangered (EN), Vulnerable (VU), Near Threatened (NT), Least Concern (LC), Data Deficient (DD)** or **Not Evaluated (NE)**.

The only British or Irish species recognized in the European Red List of Dragonflies (Kalkman *et al.*, 2010) as of 'conservation

SPECIES		Red Lists				NS	HD Annex		Domestic legislation				BAP		
		EX	EN	VU	NT	B	II	IV	WCA	WNE	S41	S42	UK	NI	Sc
Damselflies	Scarce Emerald Damselfly				G/I										
	Small Red Damselfly					G									
	Southern Damselfly		G		E		✓	✓		✓	✓	✓			
	Northern Damselfly		G												✓
	Irish Damselfly			I				✓					✓		
	Dainty Damselfly	G													
	Variable Damselfly			G											
	Norfolk Damselfly	G													
	Scarce Blue-tailed Damselfly			G											
Dragonflies	Azure Hawker			G											
	Norfolk Hawker		G					✓		✓	✓				
	Common Club-tail			G											
	Downy Emerald		I												
	Brilliant Emerald			G											
	Northern Emerald		I		G										
	Orange-spotted Emerald	G			E		✓								
	Scarce Chaser			G											
	White-faced Darter		G												

Key: **Red Lists** columns: G = Great Britain; I = Ireland; E = Europe. **NS** = Nationally Scarce (B – see text). **HD** = EU Habitats Directive (Annex II or IV); **WCA** = Wildlife & Countryside Act 1981; **WNE** = Wildlife and Natural Environment Act 2011; **S41** and **S42** = Sections 41 and 42 of the Natural Environment and Rural Communities Act 2006. **BAP** = Biodiversity Action Plan (UK, Northern Ireland and Scotland).

concern' are **Southern Damselfly** and **Orange-spotted Emerald,** both of which are categorized as **Near Threatened** (although the latter is now **Extinct** in Britain). Similar categorizations have been carried out for Great Britain (Daguet *et al.,* 2008) and Ireland (Nelson *et al.,* 2011).

GB Red List (Endangered)

Irish Red List (Endangered)

Four species are categorized as **Endangered** in Britain: **Southern Damselfly, Northern Damselfly, Norfolk Hawker** and **White-faced Darter,** and two in Ireland: **Downy Emerald** and **Northern Emerald.**

GB Red List (Vulnerable)

Irish Red List (Vulnerable)

Two species were categorized as **Vulnerable** in Britain: **Azure Hawker** and **Brilliant Emerald,** and one in Ireland: **Irish Damselfly.**

GB Red List (Near Threatened)

Irish Red List (Near Threatened)

In addition, six species are categorized in Britain as **Near Threatened**: **Scarce Emerald Damselfly, Variable Damselfly, Scarce Blue-tailed Damselfly, Common Club-tail, Northern Emerald** and **Scarce Chaser,** while **Scarce Emerald Damselfly** is **Near Threatened** on the Irish list.

GB Red List (Regionally Extinct)

Two species, **Norfolk Damselfly** and **Orange-spotted Emerald,** are categorized as **Extinct** in Britain. A third species, **Dainty Damselfly,** was also categorized as **Regionally Extinct** on the basis that it had been lost from Britain and not recorded since 1952, but annual breeding recommenced in 2010.

Nationally Scarce species

In Britain, species have also been assessed against criteria such as their presence in 10 km × 10 km squares (hectads) of the Ordnance Survey (OS) National Grid. Species with small populations that are not currently considered to meet the thresholds for categorization as **Endangered, Vulnerable** or **Near Threatened,** but occur in 16–100 hectads, are conventionally defined as **Nationally Scarce** (or Nationally Notable). This category is subdivided into **Nationally Scarce A** for species that occur in 16–30 hectads, and **Nationally Scarce B** if found in 31–100 hectads. One British species qualifies as **Nationally Scarce B**: **Small Red Damselfly.** The species account is therefore annotated with:

Nationally Scarce B

Small Red Damselfly (see *page 74*)

Legislation

European legislation

The **Southern Damselfly** is listed in Annex II of the European Union (EU) Habitats Directive and in Appendix II of the Council for Europe Bern Convention on the Conservation of European Wildlife and Natural Habitats. As a consequence, Member States are required to take action to protect it and its habitats. In the case of the Habitats Directive, action includes designating Special Areas of Conservation, such as the New Forest, Dorset Heaths and Preseli Mountains, for their populations of this species. Annex IV of the Habitats Directive lists **Orange-spotted Emerald** as a species in need of strict protection (but this species is now extinct in Britain). The species concerned are annotated as:

> **PROTECTED: EU legislation**

Domestic legislation

Wildlife and Countryside Act 1981 (as amended)

In Britain, two species of Dragonfly are listed in Schedule 5 of the Wildlife and Countryside Act 1981 (as amended): **Southern Damselfly** and **Norfolk Hawker**. This confers some degree of protection, including prohibiting handling without a licence. The species concerned are annotated in the species accounts as:

> **W&C Act 1981 (Sched. 5)**

Wildlife and Natural Environment Act 2011

In Northern Ireland, the **Irish Damselfly** is listed in Schedule 5 of the Wildlife and Natural Environment Act 2011, which similarly requires a licence to be obtained in order to handle the species. This is highlighted in the species account as:

> **W&NE Act (NI) 2011 (Sched. 5)**

Norfolk Hawker (see *page 114*)

Orange-spotted Emerald (see *page 170*)

The Natural Environment and Rural Communities (NERC) Act 2006

Under this Act a list of habitats and species that are important for the conservation of biodiversity in England and Wales has been produced. The Act also requires decision-makers such as public bodies to have regard to the conservation of biodiversity when carrying out their normal functions. The same two species that are protected under European and/ or domestic legislation are also covered by these sections of the NERC Act: **Southern Damselfly** (Sections 41 & 42) and **Norfolk Hawker** (Section 41). The species accounts are annotated as:

> **NERC Act 2006 (S41) or (S41 & S42)**

Biodiversity Action Plans

Under the UK's Biodiversity Action Plan (BAP) (see jncc.defra.gov.uk/page-5155), a list of priorities has been drawn up that recognizes the practical issues and actions required for the conservation of habitats and species. Priority Species identified in the UK BAP are those that are generally declining rapidly in the UK. Two species of Dragonfly are identified as UK BAP Priority Species: **Southern Damselfly** and **Norfolk Hawker**. These are highlighted in the species accounts as:

UK BAP Priority Species

In Northern Ireland, **Irish Damselfly** is listed as a BAP Priority Species. This is indicated in the species account as:

BAP Priority Species (NI)

While Biodiversity Action Plans have been drawn up at a national level for the Priority Species, Local BAPs have been produced by many local authorities and other bodies. These plans aim to deliver conservation at a local level and target a number of Dragonflies for actions such as surveys, monitoring or more practical conservation measures.

National Biodiversity list

One species of Dragonfly, **Northern Damselfly**, is included on the Scottish Biodiversity List. This is a list of species and habitats that Scottish Ministers consider to be of principal importance for biodiversity conservation in Scotland. Its aim is to help public bodies carry out their duty to further the conservation of biodiversity. The species account is annotated as:

Scottish Biodiversity List

Southern Damselfly (see *page 78*)

Irish Damselfly (see *page 82*)

Northern Damselfly (see *page 80*)

References

DAGUET, C. A., FRENCH, G. C. AND TAYLOR, P. (EDS.) 2008. *The Odonata Red Data List for Great Britain*. Species Status **11**: 1–34. JNCC, Peterborough. [Available from jncc.defra.gov.uk/pdf/pub08_speciesstatus11.pdf.]

KALKMAN, V. J., BOUDOT, J.-P., BERNARD, R., CONZE, K.-J., DE KNIJF, G., DYATLOVA, E., FERREIRA, S., JOVIC, M., OTT, J., RISERVATO, E. & SAHLÉN, G. 2010. *European Red List of Dragonflies*. Luxembourg: Publications Office of the European Union. [Available from ec.europa.eu/environment/nature/conservation/species/redlist/downloads/European_dragonflies.pdf.].

NELSON, B., RONAYNE, C., & THOMPSON, R. 2011. *Ireland Red List No.6: Damselflies & Dragonflies (Odonata)*. National Parks and Wildlife Service, Department of the Environment, Heritage and Local Government, Dublin, Ireland. [Available from npws.ie/publications/redlists/RL6.pdf.]

British Dragonfly Society

The British Dragonfly Society (BDS) was formed in 1983 to promote and encourage the study and conservation of dragonflies and their natural habitats. Through its various committees, local groups and members, the Society is the principal body overseeing the study, recording, conservation and education effort focussed on Dragonflies in Britain. The BDS manages The Dragonfly Centre at Wicken Fen, Cambridgeshire, in conjunction with the National Trust.

Matters of conservation concern are addressed through a Conservation Officer and a Dragonfly Conservation Group (DCG). Work includes advising and collaborating with a wide range of nature conservation bodies and others, producing advisory and educational material, and co-ordinating training events for members and other organisations. The DCG oversees the Dragonfly Recording Network (DRN) and *Migrant Dragonfly Project*, which gather information about species' occurrences and publish the findings (see *opposite*).

The BDS website provides a wealth of information about Dragonflies, including news items, help with identification, latest sightings, on-line reporting, research suggestions, a shop, and downloads including *Dig a Pond for Dragonflies*, *Managing Habitats for Dragonflies* and *Management Fact Files* for key species. You can join the Society online. Members receive half-yearly issues of *Dragonfly News* and the *Journal of the British Dragonfly Society*, and can attend the annual Members' and Recorders' Days and any of the dozens of field trips and indoor meetings arranged by local BDS groups and individual members.

For details of BDS officers and activities, visit the website: british-dragonflies.org.uk

Dragonfly recording and monitoring

Britain and Ireland have a long and enviable history of biological recording. This has been firmly based on harnessing the enthusiasm and expertise of volunteers, such as Dragonfly recorders, who can play a major role in converting their field observations into scientific knowledge and conservation action. The publication of a second Dragonfly distribution atlas for Britain and Ireland in 2014 is a major tribute to volunteer recording effort. The information from over a million records was used to produce the atlas maps, but the records hold much more information than just where Dragonflies have been seen. For example, the dates of the sightings can be analysed to determine flight periods and changes in range.

Dragonfly recording in Britain takes place under the umbrella of the Dragonfly Recording Network (DRN), which encompasses record collators and computer systems as well as the stored data. A magazine for dragonfly recorders, *Darter*, is produced annually and distributed to all recorders; it can be downloaded from the British Dragonfly Society (BDS) website. The BDS employs an officer to co-ordinate the DRN and to ensure the transfer of data to the *National Biodiversity Network* (data.nbn.org.uk/). In Ireland, recording has developed under *DragonflyIreland* (habitas.org.uk/dragonflyireland). All Irish records can be submitted online through the National Biodiversity Data Centre (biodiversityireland.ie). In Northern Ireland, they should be submitted through the Centre for Environmental Data and Recording (CEDaR) (nmni.com/cedar).

The BDS *Migrant Dragonfly Project* monitors Dragonfly migration in Britain. Immigration can often be detected by sightings of either obvious movements, unusual numbers of individuals, or Dragonflies in atypical habitats. Typical migrant species include common residents, such as Migrant Hawker and Common Darter, erratic immigrants such as Yellow-winged Darter and rare visitors such as Vagrant Emperor. In recent years, several species have been recorded for the first time in Britain and Ireland: for example, the first American species, Common Green Darner, was found in 1998, and Southern Emerald and Common Winter Damselflies were found for the first time in 2002 and 2008 respectively. Observers are encouraged to submit records of all migrants and rarities via the BDS website. Written and/or photographic evidence should support claims of rare migrants and vagrants, records of which are assessed by the BDS *Odonata Records Committee*. Details of accepted records and migration summaries are published regularly in the BDS publications, *British Wildlife* and *Atropos* (see *page 218*).

Anyone can add to the Dragonfly dataset and thereby enhance our knowledge of Dragonflies. The basic information needed to make a 'record' must answer four simple questions: What species? Where was it seen? When was it seen? Who saw it? Extra value can be added to the record if a count or estimate of numbers is made. Even more valuable is the addition of breeding information, which is currently attached to only 18% of the DRN records. All of this information can be submitted to the DRN either online or through local record collators ('Vice County Recorders'). Details of how to submit records are given on the BDS website, under the 'Recording' tab. Dragonfly records are also gathered by other organisations and passed on to the DRN. For example, the British Trust for Ornithology's *BirdTrack* and *Garden BirdWatch* include the collection of Dragonfly records. Important elements of this book, such as the distribution maps and flight-period charts, rely on information from volunteer recorders, so do help to enhance our collective knowledge by noting down what you see and submitting your records.

To monitor Dragonflies at a particular site, it is important that a standardised, repeatable method is used for counting or recording presence. Repeating this along a fixed route, at fixed times and in optimal weather can allow comparisons to be made between years and can provide an indication of how species are faring at particular sites. More broadly, a surveillance method is under development that should enable national population trends to be produced from 'complete lists' of species seen at sites on dated visits. This and other recording initiatives fall within the BDS *DragonflyWatch - Looking out for Britain's Dragonflies* framework, more details of which can be found on the BDS website.

Further reading

ASKEW, R. R. 1988 (2004). *The Dragonflies of Europe*. Harley Books, Colchester.

BROOKS, S. (ED.) 1997 (2002). *Field Guide to the Dragonflies and Damselflies of Great Britain and Ireland*. British Wildlife Publishing, Hook.

BROOKS, S. 2003. *Dragonflies*. Natural History Museum, London.

CHAM, S. 2012. *Field Guide to the larvae and exuviae of British Dragonflies: Damselflies (Zygoptera) and Dragonflies (Anisoptera)*. British Dragonfly Society, Peterborough.

CHAM, S., NELSON, B., PRENTICE, S., SMALLSHIRE, D. & TAYLOR, P. (EDS.) [in prep.]. *Atlas of Dragonflies in Britain and Ireland*. BDS, Peterborough & CEH Wallingford.

CHANDLER, D & CHAM, S. 2013. *Dragonfly*. New Holland Publishers, London.

CORBET, P. S. 1999 (2004). *Dragonflies: Behaviour and Ecology of Odonata*. Harley Books, Colchester.

CORBET, P. S. & BROOKS, S. 2008. *Dragonflies*. New Naturalist Series. Harper Collins, London.

DIJKSTRA, K-D. B. & LEWINGTON, R. 2006. *Field Guide to the Dragonflies of Britain and Europe*. British Wildlife Publishing, Gillingham.

DUDLEY, S., DUDLEY, C. & MACKAY, A. 2007. *Watching British Dragonflies*. Subbuteo Natural History Books, Shrewsbury.

GRAND, D. & BOUDOT, J-P. 2006. *Les Libellules de France, Belgique et Luxembourg*. Collection Parthénope. Biotope, Mèze, France.

MERRITT, R., MOORE, N.W. & EVERSHAM, B.C. 1996. *Atlas of the Dragonflies of Britain and Ireland*. HMSO, London. [Available from nora.nerc.ac.uk/7785/1/Dragonflies.pdf.]

MILLER, P.L. 1987 (1995). *Dragonflies*. Naturalists' Handbooks 7. Richmond Publishing, Slough.

NELSON, B. & THOMPSON, R. 2004. *The Natural History of Ireland's Dragonflies*. Ulster Museum, Belfast.

PARR, A.J. 2011. Records of exotic odonata in Britain during 2010. *Atropos* **41**: 39–42

POWELL, D. 1999. *Guide to the Dragonflies of Great Britain*. Arlequin Press, Chelmsford.

SISLBY, J. 2001. *Dragonflies of the World*. The Natural History Museum, London.

Journals

Atropos – independent journal devoted to the Lepidoptera and Odonata of the British Isles. Published three times a year. Available from: 36 Tinker Lane, Meltham, Holmfirth, West Yorkshire, HD9 4EX. Website: atroposuk.co.uk

British Wildlife – independent magazine covering all aspects of British natural history and conservation, including regular Dragonfly reports. Published bi-monthly. Available from: The Old Dairy, Milton on Stour, Gillingham, Dorset, SP8 5PX. Website: britishwildlife.com

Journal of the British Dragonfly Society – scientific papers, principally of relevance to British dragonfly species. Published twice a year. Free to BDS members (see *page 216*).

IT Resource

SMALLSHIRE, D. & SWASH, A**.** *Dragonflies & Damselflies of Britain & Ireland*. App for iPhone and iPad produced by NatureGuides Ltd. Available from itunes.apple.com/gb/app/dragonflies-damselflies-britain/id436991286

Acknowledgements and photographic credits

Many people have contributed to the production of this book and our sincere thanks go to all. It is our intention that everyone who has contributed is named in this section, but if we have inadvertently missed anyone we can only apologise. Despite the contributions of others, we hold full responsibility for any errors or omissions.

The Princeton **WILD**Guides series of field guides covering Britain's natural history is the brainchild of Rob Still, who is responsible for their design and production. Following the acclaim and success of the previous editions of this book, Rob's unrivalled – yet constantly developing – skills in computer graphics have again been used to wonderful effect in producing this fully revised third edition. Not only has Rob designed the book, including preparing the stunning plates, but he has also produced the hundreds of illustrations that feature in the adult and larval identification charts and species accounts. Without these components, this book would be nowhere near as comprehensive and we owe Rob our most grateful thanks for the countless hours he has spent seeing it to fruition.

We would also like to express our gratitude to our friends from the British Dragonfly Society for their help, guidance and support during the production of this edition of the book. John & Gill Brook, Steve Cham, Dave Chelmick, James Lowen, Adrian Parr, Steve Prentice and Pam Taylor all made valuable contributions, as did all the recorders who contributed to the DRN database. A number of other people made constructive comments on the first and second editions, which we have endeavoured to incorporate here. We are again indebted to Nick Baker for writing the Foreword and to Brian Clews for his painstaking work in checking the final draft.

The production of this book would not have been possible without the help and co-operation of the many photographers whose images have been reproduced. The plates are one of the key features of the book and we would like to acknowledge the skill and patience of all the photographers who kindly allowed us to use their work. Every photograph is listed in this section, together with the photographer's initials: John C. Abbott [JA], John Bebbington (FRPS) [JB], Tim Beynon [TB], Magnus Billqvist [MB], Allan Brandon [AB], Colin Carver (Windrush Photographs) [CC], Steve Cham [SC], Roger & Liz Charlwood (WorldWildlifeImages.com) [RC], Dave Cottridge [DC], K. Dean Edwards [KDE], Sean Edwards [SE], Jacob Everitt [JE], Mike Frost [MF], Bill Furse [BF], Robert Geerts (odonata.eu) [RG], Ted Griffiths [TG], Andy Harmer (andyharmer.com) [AH], Paul Harrison [PH], Bart Heirweg [BHe], Barry Hilling [BHi], Ian Johnson [IJ], Sami Karjalainen (korento.net) [SK], Rene Krekels [RKr], Thomas Kirchen (makro-tom.de) [TK], Tommi Laurinsalo [TL], Andrew Lawson [AL], René Manger (dutchdragonflies.eu) [RM], Andy McGeeney [AM], Simon Mitchell [SM], Joachim Müller [JM], Jochen M Müller (Libellen. Jochen.de) [JMM], Harm Niesen [HN], Ronald O'Mahony [RO], Erland Nielsen [EN], Dennis Paulson [DP], Alan Petty (Windrush Photographs) [AP], Petro Pynnönen [PP], Richard Revels (Windrush Photographs) [RR], Dave Sadler [DSa], Chris Schenk (Windrush Photographs) [CS], Peter Schütz [PS], David Sewell (Windrush Photographs) [DSe], Dave Smallshire [DSm], Andy & Gill Swash (WorldWildlifeImages.com) [AS], Tihomir Stephanov [TSt], Tim Sykes (Environment Agency) [TS], David Tipling (Windrush Photographs) [DT], Jukka Toivanen (picasaweb.google.com/jukkatt) [JT], Antoine van der Heijden (fly.to/dragonflies) [AvH], Johannes van Donge (diginature.nl) [JvD], Peter J. Wilson [PJW] and Olaf Wolfram (digimakro.de) [OW].

This third edition includes many new illustrations and images, which we believe will make it easier than before to arrive at a correct identification. We continue to express our gratitude to Jukka Toivanen, Robert Geerts and Thomas Kirchen, who generously allowed ready access to their superb images, many of which grace the stunning plates.

Last but not least, we would like to give special thanks for the patience and help given by our wives, Sue and Gill. Indeed, Dave would also like to acknowledge the part that the first edition of this book played in bringing him and Sue together!

103 **Azure Hawker:** Male (side) AM; Female (oblique) SK; Male (above) & Female (above) DSa; Female (above) SM.

105 **Common Hawker:** Female – blue (side) AS; Female – blue (above) TB; Female (above) BF; Female (above) AS; Male (oblique) and (side) AS.

107 **Migrant Hawker:** Male (side) JT; Male (above) RG; Female (oblique) IJ; Female (above) JT.

109 **Southern Migrant Hawker:** All photos RG.

111 **Southern Hawker:** Male (side) DSa; Male (above) DSa; Female (side) TK; Female (above) DSe.

113 **Brown Hawker:** Female (side) JvD; Female (above) RR; Male (side) TK; Male (above) CC.

115 **Norfolk Hawker:** Female (side) PJW; Female (above) PJW; Male (side) RG; Male (above) RG.

117 **Emperor Dragonfly** Male (above) JT; Female – blue (above) JT; Female (side) PJW; Female (above) RR.

119 **Lesser Emperor:** Male (above) RG; Female, blue (oblique) JT; Pair in copula TK.

121 **Golden-ringed Dragonfly:** Female; imm; (side) TK; Female (oblique) JT; Male (side) & (oblique) RG.

123 **Common Club-tail:** Male (above) JT; Male (side) JT; Female (above) RG; Female (side) RG.

125 **Downy Emerald:** Male (above) TK; Male imm.; (above) JT; Female (above) JvD; Female teneral (side) RG; Head (inset) DSm.

127 **Brilliant Emerald:** Female (above) AM; Female imm.; (side) SK; Male (side) DSa; Male (oblique) DSa; Male (side) DSm; Head (inset) DSm.

129 **Northern Emerald:** Female (oblique) DSa; Female (above) JvD; Male (above) BHe; Male (side) AS; Head (inset) SM.

131 **Four-spotted Chaser:** Male *praenubila* (above) RG; Male (oblique) JT; Female *praenubila* (above) RG; Female *praenubila* (oblique) RG.

133 **Broad-bodied Chaser:** Female (above) JT; Male (above) JT; Male (oblique) AS; Male imm.; (oblique) DSa.

135 **Scarce Chaser:** Male (above) RG; Male imm.; (side) JT; Female (above) AS; Female imm.; (above) RG.

137 **Black-tailed Skimmer:** Male (side) JT; Male (above) RG; Male imm.; (oblique)RG; Female (above) JT; Female (side) JT.

139 **Keeled Skimmer:** Male (oblique) RG; Male (above) RG; Female (oblique) AS; Female (above) AS; Female (above) PJW.

141 **White-faced Darter:** Male (oblique) RG; Male (above) RG; Female teneral (above) TL; Female old (above) RG; Female (above) RG; Head (inset) RG.

143 **Black Darter:** Male (side) TK; Male – obelisk OW; Female old (side) OW; Female teneral (above) RG; Female (side) JT; Male imm.; (side) BHi; Male (above) PJW.

145 **Common Darter:** Male (oblique – top left) AS; Male (above) RM; Female (side) PH; Female teneral (above) CS; Female (side) DSa; Female old (above) AS; Female old (side) AP.

146 **Ruddy Darter:** Male (side – top left) JT; Male (oblique) PJW; Male imm.; (side) RG; Male imm.; (above) BHi; Male imm.; (side – abdomen raised) OW; Female (above) AS; Female (oblique) DSa.

149 **Red-veined Darter:** Male (oblique) DSm; Male – obelisk IJ; Female old (oblique) DSm; Male imm.; (above) AS; Female (oblique) AvH; Female (above) AM; Female (side – bottom left) AS.

151 **Yellow-winged Darter:** Female (side – top left) TK; Male (side) TK; Female imm.; (side) JT; Male (above) OW; Pair in copula JT; Female, Imm; (above) RG.

152 **Blue Dasher:** DP

155 **Small Spreadwing:** All photographs DSm.

157 **Common Winter Damselfly:** Female (side) TK; Female (oblique) IJ; Male (oblique) RG; Male (side) DA.

159 **Norfolk Damselfly:** Male (side) SK; Male (side) SK; Female (side) JM; Pair in tandem SK; Male (oblique) JT.

161 **Blue-eye:** Male (oblique – wings open) AS; Male (above) AS; Male (side) DSm; Male (oblique – wings closed) IJ; Female (oblique) DSm; Female (above) IJ.

163 **Vagrant Emperor:** Male DSm; Female (oblique) HN; Female (side) IJ.

165 **Common Green Darner:** Male MF; Female TG.

167 **River Clubtail:** Male (oblique) DSm; Pair in copula PS; Female (side) TK.

169 **Western Clubtail:** Male (above) AS; Male (side) AS; Female (above) DSm.

171 **Orange-spotted Emerald:** Female (above) DSm; Male (above) JMM; Male teneral (side) DSm.

173 **Southern Skimmer:** Male (side) JT; Female (side) JMM; Male (above) JT; Female (above) RG.

175 **Yellow-spotted Whiteface:** Male (above) PP; Male (oblique) JT; Female (above) MB; Female (oblique) JT; Female on water-lily PP; Male side JvD.

177 **Ruby Whiteface:** Male (above) JvD; Female (above) JT.

179 **Banded Darter:** Male (above) RG; Male (oblique) IJ Female IJ.

181 **Vagrant Darter:** Male (above) AM; Male (side) JT; Female (above) AM; Female (side) JT.

183 **Southern Darter:** Female; old (side) JMM; Male (side) JMM; Female (oblique) AvH; Male (oblique) HN; Head & thorax (inset) DSm.

185 **Scarlet Darter:** Male (above) & (side) RG; Male imm.; (oblique) IJ; Female (above) IJ; Female (oblique) DSa.

187 **Wandering Glider:** Male (above) JA; Female (oblique) DSm; Imm; (oblique) KDE.

189 INTRODUCED EXOTICS **Common Bluetail:** Immature female AL; female DSm; **Ornate Coraltail:** DSm; **Variable Dancer:** (both) DP; **Fragile Forktail:** (both) DP; **Pale-spotted Emperor:** EN; **Slender Skimmer:** AB; **Oriental Scarlet:** DSm; **Common Redbolt:** EN.

191 The authors: AS.

194 **Large Red Damselfly:** larva RKr.

195 **Emperor Dragonfly:** larva DSa.

196 **Banded Demoiselle:** larva AH; **Scarce Emerald Damselfly:** larva RKr.

197 **Scarce Blue-tailed Damselfly:** larva RR; **Azure Damselfly:** larva RR; **Red-eyed Damselfly:** larva JT; **Large Red Damselfly:** larva RKr; **White-legged Damselfly:** larva SC.

198 **Hairy Dragonfly:** larva RKr; **Brown Hawker:** larva AH; **Emperor Dragonfly:** larva AH.

200 **Common Club-tail:** larva RKr; **Golden-ringed Dragonfly:** larva SC.

201 **Downy Emerald:** larva JT; **Four-spotted Chaser:** larva JT; **Common Darter:** larva SC.

208 **Downy Emerald:** larva (both images) JT.

209 **Four-spotted Chaser:** larva (both images) JT.

213 **Small Red Damselfly** AS.

214 **Norfolk Hawker** AH; **Orange-spotted Emerald** AS.

215 **Southern Damselfly** AS; **Irish Damselfly** JvD; **Northern Damselfly** AS.

216 BDS cap and **Common Darter** DSm.

Inside Back Cover **Small Red Damselfly** JvD; **Common Blue Damselfly** TK; **Emerald Damselfly** JT; **Emperor Dragonfly** RR; **Northern Emerald** JvD; **Banded Demoiselle** RG; **Broad-bodied Chaser** JT; **Ruddy Darter** RG.

Index of English and scientific names

This index includes the English and *scientific* names of all the Dragonflies included in this book. **Bold** names are the preferred English names used throughout. Normal text is used for alternative names or names used in Ireland or Europe; these are cross-referenced to the preferred English name, against which all the key entries for the species are listed. Page references given against scientific names are to the main species accounts, larval identification chart and selected entries.

Bold red numbers highlight the main species accounts.
Bold black numbers refer to the tables that summarise the key adult identification features.
Blue numbers relate to larvae, those in **bold** indicating the relevant page in the identification chart.
Italicised numbers indicate page(s) on which a photograph of an *adult* or a *larva* may be found.
Other numbers highlight page(s) where other key information is given.